History of Technology

History of Technology

Ninth Annual Volume, 1984

Edited by

Norman Smith

Imperial College, London

MANSELL PUBLISHING LIMITED

London and New York

First published 1984 by Mansell Publishing Limited
(A subsidiary of The H.W. Wilson Company)
6 All Saints Street, London N1 9RL, England
950 University Avenue, Bronx, New York 10452, U.S.A.

British Library Cataloguing in Publication Data

History of technology.—9th annual volume, 1984
   1. Technology—History—Periodicals
   609   T15

   ISBN 0-7201-1709-7
   ISSN 0307-5451

Printed in Great Britain by
Butler & Tanner Ltd Frome and London

Contents

Preface

This ninth collection of papers on the history of engineering and technology represents once more the wide variety of subjects and historical periods which has always been *History of Technology*'s objective.

Following last year's paper on the Roman water-mill at Barbegal it is most pleasing in this year's volume to present an English version, albeit on a different scale, of the same technology. Professor Boyer's account of pile-driving adds a further dimension to the picture of medieval bridge technology which has featured in earlier volumes. Tracts on geography and travel always have potential as sources for engineering history and Dr Donald Hill's contribution is particularly interesting because it deals with a selection of Muslim writers. José García-Diego's exploration of the life and work of Giovanni Sitoni is an important paper I believe for the light it throws on aspects of late sixteenth-century technology in Spain in general and its literature in particular.

The other three papers offered this year all relate to the nineteenth century and later. Peter Bardell's is a study of the intricate problem of the origins of alloy steels; Ian Winship deals with an aspect of agricultural history which is too often overlooked; and in Michael Duffy's paper will be found a thorough analysis of a key problem in the history of railway engineering, a subject which in content and style is not to be confused with the history of railways, although it frequently is.

NORMAN A.F. SMITH

The Origins of Alloy Steels

P. S. BARDELL

Introduction

Many of the machines and structures upon which we now depend and take for granted would not and, indeed, could not exist but for the availability of alloy steels. Despite their fundamental importance as constructional materials during the present century, they have not received much attention from the historian of technology. Often, writers commenting on the design or construction of an artifact tend to ignore or summarily and superficially pass over the materials of which it was manufactured, notwithstanding the fact that its very existence depended upon those materials possessing special, and deliberately endowed, characteristics. Materials, as opposed to contrivances, have received scant consideration. Even Volume VI of the authoritative *History of Technology* (Oxford UP, 1978), in a chapter on 'Iron and Steel', devotes but three short paragraphs to the alloys of steel.

This brief survey is an attempt, within a limited compass and based upon the source materials accessible, to at least partially rectify this deficiency and to illustrate how a number of twentieth-century engineering achievements depended upon the development and production of suitable steel alloys, and how the availability of alloy steels triggered other technological advances.

In seeking to understand and rationalize the methodology of the metallurgists who evolved new steels, it must be recognized and acknowledged that they were the heirs of a tradition whereby a corpus of knowledge had grown because earlier workers had purposely alloyed, or attempted to alloy, various elements with steel to determine the effects of such inclusions. For instance, should it now appear puzzling why an engineer at the end of the nineteenth century should even have contemplated adding a rare, expensive and difficultly obtained element such as vanadium to steel, the explanation is that he almost certainly knew that it enhanced the toughness of the steel. This fact had been known for over sixty years. In 1831, while investigating why some batches of iron were brittle while others were not, N.G. Sefström, a Swedish chemist and teacher at a school of mines, discovered in the non-brittle samples a new element and named it vanadium, after a Norse goddess.[1] Thus, from the instant of its discovery, its association with iron was known to give a better quality metal free from brittleness.

Over the years of the nineteenth century from Faraday's investigations onwards, the amount of metallurgical knowledge increased steadily so that by the time ordinary carbon steels of consistent quality could be produced in large quantities the potential for an alloy steel industry had already been established. The process of methodical examination leading to discovery provided not only new knowledge but also the means by which it could be extended. And it was. As will be seen later this procedure was used at the end of the nineteenth century by Hadfield and other workers in the field. Once it was known that manganese imparted work-hardening characteristics, chromium increased corrosion resistance, silicon reduced hysteresis losses, and so on, these alloy systems could be, and were, thoroughly investigated and developed through modifications of composition and processing in the endeavour to secure desired properties. Processing was just as important as composition. Contemporaneously, the work of F. Osmond, particularly his ideas on the theme of metallurgical phenomena being explainable in terms of structure and energy, and the extensive use of the microscope in the study of steels, led to an appreciation of the role of structure and hence of heat treatment as a major determinant of properties.

Hence the idea emerged that alloys could be 'designed' for specific purposes and this in turn had a profound influence on design philosophy. No longer was it necessary for designs to conform to the characteristics of available materials: the material could be produced to suit the design. Whereas Robert Stephenson's design of the Britannia Bridge over the Menai Straits (1850) was largely determined by the properties of wrought iron, a quarter of a century later, across the Atlantic, James Eads demanded and got the chrome steel he required to accomplish his design for the St Louis Bridge (1874).

The transition in alloy steel production and use was fundamental and extreme, too. On the one hand, interesting alloys appeared for which there was no apparent or immediate use while, on the other hand, intense and almost desperate activity, sometimes on an international scale, was initiated to produce a material to fulfil a particular need. Thus, Hadfield's manganese steel, the wonder metal with strange properties, appeared in 1882 and had to wait a decade before a significant use was found for it. Likewise, Brearley's stainless steel for cutlery was not adopted for this purpose for some seven years. At the other end of the spectrum, so paramount was the need for a material with heat-resisting and corrosion-resisting properties specifically suited to steam turbine applications that the collaborative effort of three major industrial countries was applied to the problem.

The First Industrial Alloy Steel

Plain carbon steel cutting tools, which begin to soften below 200°C, have to be re-hardened at frequent intervals by quenching in water with the consequent risk of cracking and therefore their use is ideally restricted to light cuts and moderate feed rates to avoid the generation of heat due to friction between the tool and the workpiece.

The early decades of the nineteenth century saw a rapid increase in metal-cutting operations, largely attributable to growing mechanization in the textile industries and the development of the railways which, by mid-century, was of phenomenal and unprecedented proportions extending throughout Europe as well as across the Atlantic Ocean to North America![2] These circumstances made apparent the desirability if not the necessity for improved cutting-tool metals. Until this time, tools were made by melting blister steel in a crucible from which it was poured to produce a homogeneous ingot free from the inclusions of slag streaks which then marred shear steel. Devised by the Doncaster clockmaker, Benjamin Huntsman, in 1740, this 'crucible steel' process was generally used for the production of tool steels until the advent of the electric arc furnace and the high-frequency induction furnace and, indeed, with a few old-established firms in Sheffield, persisted to the middle of the twentieth century.[3] And then, in 1855, the first alloy tool steel was produced commercially; this was the tungsten steel of Franz Köller.[4] However, although the appearance of alloy tool steels was of some consequence in production engineering, this early date can hardly be accepted as the beginning of the 'Age of Alloy Steels'. Cutting-tool steels, possessing the properties of hardenability and wear resistance, were only ever produced in very small quantities compared with the structural alloys, with enhanced strength, which came into use about a quarter of a century later. A little later than Köller's work, in 1861 when he was investigating the value of manganese in the production of Bessemer steel, Robert Mushet discovered that one composition hardened in air. Subsequent analysis showed the presence of about 6 per cent tungsten. He saw the significance of this discovery and several years later (1868) patented what is now acknowledged to have been the first air-hardening tool steel with a composition of about 2·3 per cent carbon, 2·57 per cent manganese, 1·15 per cent silicon, 1·15 per cent chromium and 6·62 per cent tungsten.[5] The alloy was remarkable in that it did not require quenching in liquid and, compared with ordinary high carbon steel, it permitted an increase in cutting speed of the order of 50 per cent. Brittleness, due to the very high carbon content (by present-day classification the alloy would be regarded as a cast iron), limited its applications at first but with the assistance of Messrs Samuel Osborne, who had been making crucible and cementation steel in Sheffield since 1852, the material was developed and improved. It then quickly acquired a pre-eminent position as a hard metal-cutting tool steel which it retained until the end of the century, when further outstanding advances were made in the development of tool alloys. Unquestionably, the work of Mushet established the value of alloy steels during the nineteenth century, although he was not the first by any means to investigate the properties of these alloys.

Early Work

Michael Faraday devoted six years' work (1819-24) to research on alloy steels with the object of improving the cheap and abundant iron, especially for the production of better quality cutting instruments. The transformation of iron to steel by the addition of carbon suggested the possibility of further improvements by the addition of other elements and with James Stodart, a cutler and maker of surgeon's instruments, he embarked on a systematic series of experiments to produce alloys using twenty different substances, ranging from gold to meteoric iron.[6] At that period, Faraday was very interested in manganese and, in 1819, read a paper on the 'Separation of Manganese from Iron' before the Royal Institution. It is therefore perhaps surprising that this element did not feature in his experiments on alloys.

Some of these experiments were unsuccessful such as, for example, the attempt to alloy titanium with steel where the lack of a satisfactory refractory material for the crucible prevented him from achieving the temperature necessary to reduce the oxide of titanium to the metallic state, with the consequence that the steel did not contain any titanium. Others, although successful, were impracticable, thus the alloys of steel with rhodium 'are perhaps the most valuable of all; but these, however desirable, can never, owing to the scarcity of the metal, be brought into very general use'.[7] Regrettably, at least from our point of view, his experiments on alloys of steel and chromium were never completed. This work of Faraday, then, which was not pursued after 1824 and did not lead to any large-scale use of alloy steels, may justly be regarded as the beginning of systematic research on alloy steels. Importantly, it aroused interest all over Europe, and Faraday's and Stodart's experiments were repeated from France in the west to Russia in the east. Activity in France was particularly intense, and under the auspices of the 'Société d'Encouragement pour l'Industrie Nationale' over three hundred experiments were carried out on alloys of iron with most of the other metals, but despite some excellent results all this endeavour did not lead to any production of alloys on a commercial basis. The ability to produce some good quality specimens on a small scale under laboratory conditions was no guarantee of success on an industrial scale and many developments in the methods, materials and general knowledge of ferrous metallurgy were necessary before steel alloys could be made and used satisfactorily. Quite apart from the impossibility of producing alloys of consistent quality during the 1820s and early 1830s, there was not a need for them really and so this early experimental work was very much ahead of its time. Indeed, commercial alloy steels are a phenomenon of the twentieth century although an intensive investigation into their properties and production was undertaken in both Europe and North America from the 1870s onwards. While a few alloys were produced on a small-scale commercial basis during the last quarter of the nineteenth century, this was essentially a period of preparation so far as the exploitation of steel alloys is concerned.

Steel Production

The development of useful alloys obviously depended upon the existence of the right conditions, and paramount amongst these was the ability to produce, in large quantities, a complete range of carbon steels of consistent quality which, in turn, required an adequate knowledge of chemistry and metallurgical science together with the necessary production and testing techniques. Furthermore, an available supply of alloying elements at an economic price was also essential, together with a demand for the alloys produced. This was provided, initially, by military needs and then, particularly in the USA, for structural engineering purposes, but undoubtedly it was the appearance of the petrol engine, the steam turbine and the rapidly-growing infant electrical industry which provided by far the greatest impetus to the production of alloys possessing specific and constantly improving properties.

Large-scale steel production is generally acknowledged to have begun with Bessemer's spectacular experiment at his bronze factory at St Pancras, in 1856.[8] But several years' hard work were necessary to overcome the formidable difficulties involved in making his process industrially viable.

His initial success, attributable to the fortuitous choice of both a phosphorus-free Blaenavon grey iron as his raw material and a clay or firebrick lining for the converter, was not shared by his first licensees who found the metal poured from the converters to be brittle and full of blowholes. The brittleness was due to the high phosphorus content of the pig iron; in excess of about 1 per cent, this element made the steel too brittle for the majority of industrial uses. And since it was present in the majority of native and European iron ores, it provided both a difficulty and a challenge to steel-makers. For well over a decade the problem of eliminating phosphorus confounded the 'Bessemer' operators. The users of the open-hearth process were likewise troubled.

Thus, it was not until the 1870s that large-scale steel production began and even then it was generally based on the use of non-phosphoric ores, and it was at this period that serious investigations of the effects of alloying elements started.

Persevering collaboration between the cousins S.G. Thomas and P.C. Gilchrist succeeded in resolving the intractability of phosphorus, and so by the 1880s the world's phosphoric iron ores were amenable to industrial exploitation and the stage was set for the appearance of alloy steels.

Ferro-Alloys

Alloy steels may be made either by making the necessary additions during the steel-making process itself or, in some cases, there may be a separate alloying stage. The additions may be made in the elementary form, which is the case with copper and nickel, but other elements such as silicon, manganese, chromium, tungsten, molybdenum and vanadium are usually added as ferro-alloys, that is, alloyed with iron and containing carbon as well, which are easy to obtain directly by smelting

In a paper published in the *Annales de Chimie* (Vol. XVII, 1821) on the alloys of chromium with iron and steel, P. Berthier described the preparation of ferro-chromium which he thought would be a convenient method of adding chromium to cast steel.[9] He is usually acknowledged as being the first to recognize the importance of ferro-alloys in the manufacture of steel.

Metallurgists at the Terre Noire Steel Co., who later carried out systematic researches into the effects of alloying elements on iron, made use of ferro-alloys but their results were vitiated by the inconsistency of these due either to being too highly carbonized or deficient in the appropriate element. One of this group of metallurgists, A. Pourcel, succeeded in producing ferro-manganese with the blast furnace, which proved to be an achievement of the utmost importance for the production of steel since it effected an immediate reduction in the cost of this ferro-alloy, used as a deoxidiser, from about £56 per ton to £16 per ton.[10] As techniques advanced, the quality of ferro-alloys improved, especially during the early years of the twentieth century when the electric arc furnace, which was essential for those requiring a high temperature for their formation, came into use. Electric ferro-silicons and 60 per cent ferro-chrome were first marketed in Britain by the Liverpool firm of George Blackwell, Sons & Co., who introduced a number of new metals and alloys to steel-makers including, by 1913, rarer products such as ferro-uranium and tantalum.[11]

Defence and Offence

The idea of protecting ships by a metal casing dates back to at least 1835 when Stanislas C. Dupuy de Lome planned to encase, down to the water-line, the propeller-driven ship *Napoleon*.[12] His scheme did not come to fruition, but by the 1850s armour plate was being used for this purpose, although there is uncertainty as to its original application. According to Paul G. Bastien the first use, in 1854, was on the floating batteries *Lave*, *Tonnante* and *Devastation* which were protected by 110 mm iron plates.[13] In the early years of this same decade, Messrs Samuel Beale and Company of the Park Gate Works had also rolled armour plates, 13 ft × 3 ft × $4\frac{1}{2}$ inches thick and weighing 72 cwt, for the early iron-clads. At a cost of £250,000, John Brown's built an armour-plate mill in 1863 capable of rolling plates 12 inches thick and weighing 20 tons. Even as late as 1873, wrought iron plates 10 inches thick were being made by Charles Cammell & Co. at Sheffield.[14] However, as the penetrating power of shells increased, steel replaced iron for this purpose and tests at Spezzia in 1876 conclusively demonstrated its superiority, but hardness, a desirable property for armour plate, was obtained in steel by increasing its carbon content which, in turn, increased its brittleness which was undesirable, and this situation led to attempts to improve the steels by alloying.

Nickel Steels

Following the work of Henri Marbeau during the late 1870s and early 1880s on nickel steels, James Riley, who saw these alloys produced in France during September 1888, carried out investigations of his own and presented his findings to the Iron and Steel Institute in May 1889. His steel, containing 4·7 per cent nickel, showed considerably greater strength than carbon-steel without any significant loss of ductility.[15] Alloys of this composition were also developed at Le Creusot, while at the same time the Montluçon Works were working on chromium-steel plate. Ballistic tests carried out by the United States Navy at Annapolis in 1890 showed the nickel steels to be vastly better than the carbon steels and so convincing were the results that the French steel industry immediately adopted nickel steel, although England lagged behind in making the change until 1895. Rapid developments were then made in the manufacture of armour plate with the lead being held by France, using alloys containing 5 per cent nickel, and Germany, whose alloys were based on a 7 per cent nickel content. The penetration resistance to shells was further increased in 1892 when the Saint-Chamond Works added chromium to the nickel steel.

Improvements in the means of defence obviously stimulated the development of even more destructive missiles and at this period shells of crucible steel, forged and quenched at the nose, replaced those made of hard cast iron and of cast steel. To combat this situation, the task of the armour plate manufacturers was to produce plates with a hardened face so that upon impact the shell would fracture, and this was achieved by heat treatment, that is, by case hardening and quenching. The case carburizing was accomplished by the use of solid materials (Harvey process) or by gas (Le Creusot process) which was used for this purpose from 1892. However, the hardness was obtained at the expense of brittleness; but by increasing the nickel content and modifying the quenching treatment, the Krupp Works were able to produce a case-carburized armour plate that did not exhibit brittleness in a ballistic test. Le Creusot, in 1906, and Saint-Chamond produced a homogeneous armour plate containing nickel, chromium and molybdenum which was hardened by differential heating and spray quenching which proved to be equal to Krupp armour plate in resistance to uncapped shells and superior in resistance to capped projectiles.

The alternating advantage of armour and armament was responsible not only for great progress in improving the properties of steel, thereby promoting its acceptability to the many engineers who were somewhat distrustful of it and preferred the better-known wrought iron; it also considerably widened the experience of steel-makers and stimulated a more systematic understanding of ferrous metallurgy.

By the end of the nineteenth century, the science of metallurgy was definitely emerging as a result of both a current interest in the properties of materials for specific purposes, for example, the production of armaments just mentioned, and also the availability of instrumentation which

permitted scientific study. From the middle of the century it was known
that metals were of crystalline structure—a fact which macroscopic examination had revealed—but from 1886, when H.C. Sorby's paper 'On
the Application of Very High Powers to the Study of the Microscopical
Structure of Steel' was published,[16] the microscope was seriously applied
to this task. Josiah Willard Gibbs, in 1877, enunciated the Phase Rule
which formed the basis of organized knowledge of metal equilibria, while
in the same year Henri Louis Le Châtelier, an outstanding scientist interested in applying scientific methods to industrial problems, suggested the
use of a thermoelectric couple for measuring high temperatures. Taking
advantage of this, Floris Osmond made a study of iron-nickel alloys and
established that the effect of increasing the nickel was a progressive low-
ering of the transformation temperatures of iron; that is, the addition of
nickel acts similarly to increasing the rate of cooling of a carbon steel. He
also distinguished the difference between steels containing less than 30 per
cent nickel and those with more than this amount which leave the steel
non-magnetic, even after a slow cooling to room temperature. Such steels
were of growing importance in electrical engineering and in 1895 the
Imphy Steel Works started the commercial production of an austenitic
nickel-chromium steel (0·5–0·6 per cent C, 2–3 per cent Cr, 22–24 per
cent Ni) under the name NC4.

Charles-Edouard Guillaume, a Swiss scientist who worked at the International Bureau of Weights and Measures, and who was looking for a
suitable substitute for iridium-alloyed platinum as a secondary length
standard, noticed that this alloy had an expansion one and a half times
that of iron and nickel, while an iron-nickel alloy with 30 per cent nickel
had a lower coefficient of expansion around room temperature than that
of platinum, and became interested in this series of alloys. A systematic
study of them in cooperation with the Imphy Steel Works led, in 1897, to
the discovery of the 36 per cent nickel alloy with almost a zero coefficient
of expansion, known as 'Invar'. This alloy has remained in use to the
present day for the manufacture of pendulum rods for master clocks,
measuring tapes, parts of precision instruments and delicate sliding mechanisms for use at varying temperatures, while in a more mundane environment, marine boiler tubes of this alloy have been found to last three times
longer than carbon-steel tubes. Following further research into this
material, Guillaume produced 'Elinvar' (where Young's modulus remains
constant with variations in temperature around 20°C), an alloy, important in chronometry, which was produced industrially in 1920.[17]

Structural Steels

Although until the First World War the bulk of the nickel steel produced
was for military use, from the beginning of the twentieth century the
low-alloy nickel steels, containing up to about 5 per cent nickel, have been
increasingly used commercially and the trend continues. They are similar
to carbon steels but possess improved tensile strength and toughness

because of the finer pearlite formed and the presence of nickel in solution in the ferrite. Additionally, the grain-refining effect of nickel makes the low-nickel, low-carbon (up to about 0·1 per cent C) steels very suitable for case hardening, and they have been developed for applications such as gears, shafts, cams, crank-pins, etc., where a combination of surface hardness and core toughness are required. They, like almost every other series of alloys, have been widely used in automobile engineering.

The transition from wrought iron to steel for constructional purposes was not a swift affair although it occurred earlier in the United States, where the first complete steel-framed building was erected in Chicago in 1884, than in Europe, where, as late as 1889, the Eiffel Tower was erected with wrought iron. It was not until 1877 that the Board of Trade in Britain authorized the use of steel in bridge-building, and the Forth Bridge, built 1882–9, was the first major bridge to be constructed with this material, in fact 54,000 tons of it.

James B. Eads had a steel containing chrome produced to satisfy his requirements when he built his arched St Louis bridge over the Mississippi River between 1869 and 1874. So, too, in 1902, the design engineers of the Queensboro Bridge, which was to span the East River in New York City, requested the development of a stronger steel to enable them to confine the number and dimensions of the supporting members within the space and weight limitations of their design. Unlike earlier wrought-iron structures such as the Britannia Tubular Bridge across the Menai Straits or the Royal Albert Bridge over the River Tamar where the design was largely governed by the materials available, we see here the design demanding a new material. This situation was to characterize many engineering developments in the future. In this case the demand was met by the Carnegie Steel Company (now part of the United States Steel Company) who supplied a 3·25 per cent nickel steel which proved to be satisfactory from the point of view of strength.[18] Nickel steels were subsequently used in many large American bridges including the Manhattan and the George Washington in New York City, the Benjamin Franklin joining Camden and Philadelphia and the Bay Bridge linking San Francisco and Oakland. The addition of nickel would have raised the yield point of the steel about 50 per cent above that of mild steel and so the increased material cost was offset by the gains in terms of weight reduction and smaller-sized members.

Naturally, the search for cheaper structural steels was not neglected and from the 1880s it had been suggested that a silicon steel containing about 1 per cent silicon and 0·25 per cent carbon, a constructional material cheaper than nickel steel, could be used in shipbuilding. The complicated shapes in certain parts of a ship's structure created difficulties since steel had to be formed differently from iron which demanded new fabrication skills and techniques, but as rolling methods improved and the hammering of plate ingots gradually gave way to roll cogging the difference in costs between iron and steel narrowed and the advantage passed to steel as a constructional material. This had been further accentuated

when, in 1877, Lloyds agreed to a reduction of 20 per cent in the weight
of plates and scantlings for steel, in relation to iron, for shipbuilding and
when they first established rules for steel ships in 1888. Such silicon steels
possessing high tenacity and resistance to shock loading without loss of
ductility were used in the construction of the Cunarders *Lusitania* and
Mauretania in 1907. Despite Sir Robert Hadfield's glowing description of
the qualities of this alloy in his discussion of these two ships,[19] its use
apparently caused many difficulties and manufacture of the alloy was
discontinued.[20] Later, a so-called 'silicon structural steel'—a misnomer
since the silicon content was merely nominal and the steel depended on
a relatively high carbon content (usually 0·3 per cent minimum)—was
produced and used for the first time in 1915 in a bridge across the Ohio
River at Metropolis, Illinois. It was also used for the arch trusses of the
Sydney Harbour Bridge completed in 1932.

Another alternative to the expensive nickel steel, a 1·6 per cent manganese steel, was introduced by the American Bridge Company in 1927
when they used it for the lower chord members of the Kill van Kull
Bridge at Bayonne.

Similarly, in Europe a variety of solutions were found for the various
structural problems encountered but always, as it seems that any engineering project must, they involved compromise. All the high-strength
steels developed before 1930 lacked one or more of the qualities required
for most constructional applications. Thus the 3 per cent nickel steel
developed by the Carnegie Steel Company, mentioned above, can, with
suitable heat treatment, have its yield point raised to three times that of
mild steel—as, indeed, it does for use in engines—but it is not practicable
to quench and temper large plates and long sections and so in structural
engineering this steel has to be used in the as-rolled condition, as do many
others.

In general, it may be concluded that while alloying elements effect
some improvement in the mechanical properties of the steel without heat
treatment, the maximum enhancement is only gained after such treatment. The yield point and tensile strength, two important parameters in
structural work, are raised by small additions of manganese, nickel,
chromium and molybdenum usually in combination where their effect is
greater than if they are used singly in larger quantities.

During the 1930s and particularly throughout and since the Second
World War, considerable attention has been given to the production of
steels with good welding characteristics because this method of joining
permits a more economical use of metal with a consequent weight reduction. This necessitates a low carbon content (0·15–0·23 per cent C) to
avoid cracking in the heat-affected zone adjacent to the weld, which in
turn limits the yield points and tensile strengths attainable in weldable,
as-rolled steel. Although brittle fracture was not a new phenomenon and
certainly not confined to welded structures, the problems associated with
it were dramatically brought to prominence by the catastrophic failure of
some of the all-welded 'Liberty' ships during the war of 1939–45. Result-

ing from this unhappy experience much subsequent work has concentrated on the techniques of welding, the design of welded structures and the production of steels with a high yield point and suitable for welding. The United Steel Companies Ltd, in the period 1947–53, developed 'Fortiweld', a steel alloy of this sort containing 0·5 per cent molybdenum and 0·0013–0·0035 per cent soluble boron.[21] It is interesting to note here the use of the metalloid boron as an indication of current trends. The general effect of boron in steel is to increase the ultimate strength and the elastic limit. To obtain certain characteristics it can substitute for nickel, chromium, manganese, vanadium or molybdenum in alloy steels and the need to conserve the more costly and scarce metals points to a growing exploitation of the element.

Atmospheric corrosion confronts the engineer all the time and so besides the use of high-cost stainless steels on the one hand and the protective coatings of metal or paints on the other there has been a need for a low-alloy steel that rusted more slowly than mild steel. Atmospheric exposure tests have revealed the efficacy of very small amounts of copper, with or without the presence of chromium, in retarding the corrosion rate of mild steel, and the American alloy 'Corten' with a composition of 0·1 per cent carbon, 0·5 per cent silicon, 0·5 per cent copper, 0·8 per cent chromium, 0·1 per cent phosphorus and 0·8 per cent manganese and a corrosion rate of 40–60 per cent that of unalloyed steel has satisfactorily met this requirement.[22] Through patient and painstaking work, then, particular problems have been tackled and acceptable solutions found.

Sir Robert Hadfield and his Work

Sir Robert Hadfield's discovery of manganese steel in 1882, as the result of a series of experiments inspired by the brochure of the Terre Noire Company at the Paris Exhibition of 1878, is generally acknowledged to open the era of structural alloy steels—or at least this appears to be so in England.

It was truly a remarkable discovery; Hadfield himself calls it 'the discovery and invention of manganese steel ...' although he does not distinguish between the two processes or indicate where 'invention' contributed to the process.[23] One may be forgiven for suspecting that invention was conspicuously absent from the whole affair in so far as the steel revealed characteristics quite the opposite from those which might have been expected. Unlike carbon steel which is hardened but embrittled by quenching, his alloy was toughened and its resistance to abrasive wear increased with the severity of the service to which it was subjected. With suitable heat treatment it developed a high tensile strength (60–70 tons/in²) with considerable ductility—Hadfield quotes an elongation of 50–70 per cent on a gauge length of 8 inches. Furthermore, although the steel contained 85–87 per cent iron, water quenching rendered it non-magnetic. As exceptional as its properties, however, was the fact that there was no apparent and immediate application for this unusual alloy. The objective of his experiments was to obtain a very hard steel for 'tramway wheels, and grinding wheels and discs to be used in the place of emery wheels', but its properties rather restricted its utilization and demanded special methods for handling it. Appropriate uses had to be found for the steel and it was not until 1892 that it was used on any significant scale and then it was in the United States where William Wharton & Co. tried it, successfully, for tramway points and crossings.[24] Not until 1901 was it put to this use in England by the Sheffield Corporation Tramways for the track layout in the city's busy Fitzalan Square. Later, on the most severely tested section of a layout enlarged to accommodate increased traffic, this alloy proved to have a life forty-eight times that of the carbon steel originally used.

After the difficulties in manufacture and application had been overcome, manganese steel securely established itself as the standard material for the heavy-duty sections of both railway and tramway tracks, and the outstanding success of castings for points and crossings led to a demand for rolled manganese steel rails. In 1925, Sir Robert was able to report that experience had already shown them to have a life of five to six times that of ordinary steel rails. Throughout the passing years of the present century, this alloy has continued to find wide usage where conditions require the highest wear resistance, as for example the jaws of stone- and ore-crushing machines, the buckets and other wearing parts of dredgers and excavators, the wheels of mine cars, wire line sheaves for oilwell machinery, and so on. Hadfield cites the case of some jaw crushers produced by his own firm for use in a stone-crushing machine, weighing 100 tons, and used in the construction of the Cauvery Metur Dam in India. With a capacity to crush 200 tons of stone per hour, the machine's crushers, built entirely of cast steel, with a feed opening measuring 54 by 42 inches, were able to cope with blocks of stone weighing over 2 tons each, reducing them to 6 inch cubes in about a minute.[25] Clearly, the impact of such a machine on major civil engineering projects was tremendous.

Today, the original composition is still used although its manufacture has been improved as, indeed, have its properties by progress in heat-treatment processes. Not least among the benefits attributable to the discovery of this alloy was the stimulus it gave to Hadfield and others to investigate the properties and effects of other alloys and alloying elements. Before 1890 most of the elements now commonly used in steels, that is, aluminium, chromium, cobalt, molybdenum, nickel, tungsten and vanadium, were little more than chemical curiosities. The possibilities and potentialities of alloy steels were, however, clearly appreciated and as Hadfield later wrote:

> Following his work on manganese and silicon steels, the author has also carried out correlated studies on alloys of iron with aluminium, chromium, nickel, tungsten, cobalt, molybdenum, copper, titanium, and other elements. In almost every case these researches have covered and described methods of manufacture; chemical composition and analysis; properties of the metal as cast, and as rolled,

forged, hammered, or pressed; heat treatment; mechanical qualities, including elasticity, tenacity, elongation, and resistance to shock; hardness tests; microstructure; electrical and magnetic qualities; thermal conductivity; resistance to corrosion and erosion; and other properties where possible. In many instances specimens of the alloys as they appeared were placed at the disposal of the scientists and specialists of many countries and in various branches of scientific work, the results being of great mutual benefit.[26]

This clear statement by one who was directly involved and in fact at the forefront of the intensive research work carried out provides ample evidence of the systematic method and thorough nature of the endeavour made and, just as important, a willingness to make known and share the findings and benefits of their researches.

Electrical and Magnetic Materials

Although Hadfield had started his researches on silicon iron in 1882, it was almost twenty years before their significance and value was realized. With the thoroughness mentioned above, the electrical and magnetic properties of over a hundred alloys produced by Hadfield were measured by W. F. Barrett and W. Brown, and the results published in the Proceedings of the Royal Dublin Society, 1900, revealed that additions of silicon or aluminium effected a marked reduction in hysteresis losses while increasing the permeability and resistivity of the iron. From an electrical aspect, this was a fortunate combination of properties because the high resistivity reduced the eddy current losses. While there is disagreement on who first recognized the importance of silicon iron, and whether its first applications were in England, Germany or the United States, it cannot be doubted that with the introduction of alternating current systems of power generation in the mid-1880s, following developments of electromagnetic communications earlier in the century, there was a general search for materials with improved magnetic properties. Actually, of Hadfield's original alloys, those of aluminium iron had better magnetic properties than the silicon irons, but the difficulties caused by oxide inclusions made the commercial production of these alloys impossible at the beginning of the century and even today, with the benefit of many advances in steel-making, the aluminium irons have not become commercially important.

In the thirty years from the beginning of the century, pioneering work on the silicon irons was carried out by Hadfield in England, E. Gumlick in Germany and by T. D. Yensen in the United States. From about 1911, Yensen carried out a lot of work with vacuum furnaces and established the value of vacuum melting in obtaining materials with higher permeabilities and lower hysteresis losses. Over this period, and largely resulting from their work and that of others, the main improvements in the commercially available alloys have resulted from the elimination and control of impurities.

Improvements of greater consequence have stemmed from increasing scientific knowledge. For example, the work of P. Weiss, starting in 1907, led to the domain theory of ferromagnetism, and research on single crystals of iron and iron silicon alloys dating from 1918 showed that the magnetic properties of single crystals are anisotropic. This culminated in the work of Norman P. Goss, who established that by a combination of cold-rolling and annealing a silicon iron could be produced with outstanding magnetic properties in the direction of rolling, and on 7 August 1933 he applied for what is the basic patent on grain-oriented silicon iron.

Developments to meet the specific requirements of power generation and transmission equipment, communications equipment and electronics have led to the emergence of groups of soft magnetic materials which, for convenience, may be divided into high-permeability, high-resistance and high-saturation alloys. However, only the first group is of relevance here since the other two comprise materials quite unrelated to alloy steels. For example, the development of the oxide ferrites, which have achieved great importance as high-resistance materials in high-frequency applications, took place quite independently of the development of metallic alloys. Besides the series of iron-silicon alloys containing up to about 4·5 per cent silicon, developed from Hadfield's original alloys of this type and used for the construction of transformer cores, as laminations, and motor and generator armatures, another series of high-permeability alloys was developed based on the binary system of iron and nickel. While this system was investigated at the end of the nineteenth century by J. Hopkinson,[27] its importance was not recognized until the results of G.W. Elmen's work was published in 1916. Following the introduction of these alloys, from about 1920 onwards, ternary and more complex alloys appeared, often with properties enhanced by heat treatments such as high-temperature annealing in hydrogen or annealing in a magnetic field, but these improvements have been obtained by the progressive diminution and, in many cases, elimination of iron from the alloys.

Naturally the permanent magnet alloys, or the hard magnetic materials, also received considerable attention and, starting from quenched high-carbon (1 per cent C) steels giving the highest values of remanence and coercive force attainable with the plain carbon steels, from about 1910 the power of these magnets was increased by the addition of chromium (3-6 per cent) and tungsten (5-6 per cent). Further improvements were achieved in Japan in 1920 when K. Honda and S. Saito developed a series of alloys containing 30–50 per cent cobalt, and yet better results were obtained by T. Mishima in 1931 who produced a series of ternary alloys of iron, nickel and aluminium to which he subsequently added cobalt, in quantities varying from 0·5–40 per cent, which produced grain refinement with, again, a consequent increase of coercive force and residual magnetism. As with the soft magnetic materials, properties have been improved by the production of anisotropic alloys and by annealing in a strong magnetic field, etc. Since they are too hard and brittle to forge or machine, these alloys have to be produced as castings and ground to finished size, but compared with the plain carbon steels of the early 1900s their available magnetic energy per cubic centimetre of material has increased as much as 25–30 times.[28] This has enabled great progress to be made in many branches of electrical engineering, especially in the fields of communication and instrumentation.

Again, as with the soft magnetic materials, the trend is unmistakable: improvements have been and continue to be obtained at the expense of the iron content and it seems that the point has been reached, if not passed, where it is more accurate to refer to these materials as 'magnetic alloys' rather than as 'steel alloys'.

Corrosion- and Heat-Resisting Steels

At the turn of the century, two developments occurred which gave tremendous incentives to the development of alloy steels. One was the adoption of steam turbines in preference to reciprocating engines to drive generators and the other was the appearance of the motor car.

Between 1900 and 1938 the nominal rating of turbo-generators increased from about 5 MW to 160 MW while corresponding steam pressures rose from 140 lb/in^2 to 1,200 lb/in^2 and operating temperatures went up from 350°F to 900°F.[29] This continuing rise in operating pressures and temperatures demanded, and in turn was made possible by, steels which retained their strength and hardness—to withstand the erosive action of high-velocity steam—together with a good resistance to corrosion.

Originally, nickel steels (3–7 per cent Ni) were used, in many cases with complete success but in others trouble due to corrosion was encountered, especially through salt contamination in marine turbines. Nickel contents were increased, up to as much as 30 per cent, but high-nickel steels did not solve the problem, and bronze fared no better. It coped with the corrosion as might be expected, for this was how it had justifiably earned its high reputation in marine environments, but it suffered severely from erosion and was quite unsuitable for sustaining the effects of superheated steam. The work of H. Brearley had shown that steels containing 12–14 per cent chromium, when in the hardened condition, were successful in resisting atmospheric and most other forms of corrosion but difficulties were still presented by contaminated steam. His alloy offered the possibility of producing good turbine blades, although its great disadvantage was its dependence upon heat treatment which could be adversely affected during the assembly of the turbine which sometimes involved brazing, welding or even casting the blades in the disc. Furthermore, the blades had to possess a good surface finish because blemishes and defects acted as foci for corrosion and erosion. Eventually these difficulties were overcome by modifying the design of turbines and developing blade alloys free from these weaknesses, but it took a long time and when the alloy did win recognition and gain prominence it was through its use for the manufacture of cutlery.

Although this stainless steel, in the hardened and tempered condition, was ideally suited for the manufacture of cutlery and Brearley himself suggested this use to his employers, Messrs Thomas Firth & Sons, they were not interested in the new material and from the end of 1913 he had to engage, and persevere, in a lengthy campaign, which involved himself and Firth's in a patent dispute, to gain acceptance of his new alloy. Not until the 1920s did it actually come into use for this purpose. The corrosion resistance of Brearley's type of steel deteriorated when it was in the soft condition but, also in 1913, in Germany, B. Strauss and E. Maurer developed another stainless alloy with a much higher chromium content which was amenable to cold-working operations, such as drawing or pressing, and which possessed its maximum resistance to corrosion when in its softest condition. This steel, the famous '18-8' (18 per cent Cr, 8 per cent Ni and about 0·1 per cent C), is austenitic in its annealed state. It was not introduced into Britain until 1923.

During the 1920s, considerable effort was put into the development of corrosion-resisting steels, all depending mainly on chromium for this property, and many different alloys appeared. For convenience, they may be classified in three groups.

(1) Martensitic stainless steels, containing 0·25–0·3 per cent carbon. These are amenable to heat treatment and are generally used in the quenched and tempered condition for cutlery and other purposes where good wear-resistance properties are required.

(2) Ferritic stainless steels, with only 0·07–0·11 per cent carbon, cannot be heat treated, and were introduced in Sheffield, in 1920, in response to the demand for softer, cold-working alloys.

(3) Austenitic stainless steels, which also possess a very low carbon content, are not responsive to heat-treatment processes. They are characterized by high chromium and nickel contents and possess superior corrosion resistance to the other two categories although exposure to industrial atmospheres containing sulphur fumes, to marine environments, salt spray and various chemicals can cause pitting corrosion, which is prevented by the addition of 2–4 per cent molybdenum.

Failures of stainless steels in resisting corrosion due to phenomena such as acid attack and 'weld decay' have led to the development of many complex alloys, some containing up to six alloying additions of elements such as molybdenum, vanadium, niobium, nickel, copper, aluminium and titanium.

However, the pace-setting demands of prime movers must again be considered. As the result of the co-operation between three firms, Société Anonyme Commentry-Fourchambault et Decazeville, Messrs Hadfield's Ltd and the Midvale Steel Company of the USA, a nickel chromium steel known as ATV,[30] containing 30–36 per cent nickel, was developed for steam turbine use. Experience proved it to be non-corrodible under all the service conditions encountered—and in some continental generating stations the steam was badly contaminated with salt—and also highly resistant to erosion.[31] Additionally, it retained its mechanical strength at the highest operating temperatures while its lower density, compared with other alloys used for this purpose, gave correspondingly lower centrifugal stresses. Troubles arising from differential thermal expansion were also prevented since its coefficient of linear expansion was similar to that of the ordinary or nickel steels used for the rotors. Unlike the situation at the beginning of the century where suitable uses had to be found for manganese and silicon steels, the exacting requirements of the steam turbine here instigated a major engineering and metallurgical programme of development, carried out in three leading industrial countries, to produce an alloy for a specific task. During the 1920s the alloy was used with success for blading in the 50 MW turbine at the Gennevilliers power station, near Paris, and in the 85 MW installation of Imperial Chemical Industries Ltd, at Billingham. It was adopted by various navies of the world and at the time Sir Robert Hadfield was writing, in 1930, it had been specified for the turbine blading of the new liner *Ile-de-France*. This enormous ship, 1,017 ft long and with a beam of 115 ft, was driven by turbines developing 160,000 shp. The ability of ATV to withstand the most severe conditions ensured its use for other steam turbine fittings, nozzles, spindles, valve seatings, springs and so on, not only in applications involving high degrees of superheated steam but also in less demanding environments where its superior qualities could effect economies through prolonged life and lower maintenance costs.

The pursuit of ever-higher performances from boilers and steam plant, furnaces and the internal combustion engine, especially as applied to the aeroplane, concentrated more and more attention on the production of the so-called 'heat-resisting' steels. That is, steels which retain their mechanical strength at elevated temperatures while also resisting oxidation and corrosion. Throughout the 1920s the alloys found best able to meet these requirements were those containing high percentages of nickel and chromium, the latter element contributing largely to the retention of strength, although this diminished rapidly at temperatures above 650°C, and resistance to scaling, but at temperatures of 850°C the non-scaling characteristics disappeared too. For specific purposes, small additions of elements such as silicon and tungsten brought about improvements of the desired properties, but by this time the iron content of some of these steels was only about 50 per cent and the trend was being established of superior characteristics being achieved by increasing the alloying elements at the expense of iron. There were constant pressures to improve the power–weight ratio of aircraft engines and considerable impetus was given to the improvement of design and construction of internal combustion engines by sporting activities, in particular the pursuit of speed records on land, water and in the air, and endurance trials such as the 24-hour races at Le Mans which began in 1923. So great an influence did the Schneider

Trophy exert, for instance, that in the nine months prior to the 1931 race the Rolls-Royce 'R' engine, which powered the winning Supermarine S6B, had its rating increased by 450 hp to give an output of 2,350 hp. The experience gained by R.J. Mitchell, the Supermarine designer, was later put to good use in the design of the Spitfire, the fighter aircraft which played such a decisive role in the Battle of Britain. Sir Robert Hadfield, with justifiable pride, has written of the records won for England in the later 1920s, and the contribution of steels developed through prolonged research by his company, in collaboration with the companies referred to earlier, to these successes. Their alloy 'Era/ATV' (a high nickel chromium steel), used in the rotors of exhaust gas turbines running at speeds of 30,000 rev/min (50,000 rev/min on test), was able to withstand the high centrifugal and other stresses at operating temperatures of 800 to 930°C and, as he wrote, 'it has proved itself more than equal to the highest requirements yet imposed by the designers on the exhaust valves of motor car and aeroplane engines'.[32] To provide cheaper materials for less severe working conditions, although still able to meet the requirements of high-performance engines, Messrs Hadfield's Ltd produced the 'Hecla' series of alloys. 'ATV', 'Era' and 'Hecla' steels are still produced today. The available evidence has necessarily confined this brief survey to the activities of one firm but the same pressures and stimuli were affecting the other steel-makers, too, and comparable heat-resisting steels were manufactured by, amongst others, Messrs Firth Brown, Edgar Allen and the English Steel Corporation.

The development of gas turbine aero engines, intensified dramatically by the demands of the Second World War, made even more stringent demands for alloys able to cope with the most rigorous conditions of high stress and elevated temperatures combined with resistance to the corrosive products of combustion in the engine. Undoubtedly the most successful was the 'Nimonic' series of high-temperature alloys, developed by Messrs Henry Wiggin, based upon the nickel-chromium binary system but progressively improved by 'stiffening' with small amounts of two or more of the elements, carbon, aluminium, cobalt, molybdenum, titanium and zirconium, in suitable combinations to form small particles of carbides or intermetallic compounds which in turn raised the limiting creep-stress of the alloy. Through intensive research and test work extending over a period of more than a quarter of a century and all for one specific application, that is the gas turbine engine, almost unbelievable improvements have been accomplished. Thus 'Nimonic 80A', one of the earliest alloys in the series, had a tensile strength of 5 ton/in² at 1,000°C, while 'Nimonic 115', at the same temperature, has a tensile strength of 28 ton/in². The iron content of these two alloys is 1·5 and 0·5 per cent respectively and they cannot really be regarded as steel alloys at all, but they are mentioned because they are made by steel-makers, using steel-making methods, and also to emphasize the fact that the future for 'critical' engineering components to withstand elevated stresses and temperatures seems to lie with the refractory metals.

The beginning of the twentieth century saw the birth of the motor car industry, and this immediately presented steel-makers with problems satisfying its demands not only with regard to the construction of petrol engines, but also in the production of a chassis, suspension and transmission systems and bodywork. Vanadium was the element that was to play a vital role in the progress of the automobile, and it continues to do so today, and its story in connection with the design and production of Henry Ford's 'Model T' has already been interestingly, although somewhat briefly and superficially, recorded.[33]

Vanadium steels had been tested in France as early as 1896 and concise reports of the effectiveness of the element in increasing the tensile strength of steels appeared in the *Journal of the Iron and Steel Institute* during 1901 and 1902, but it was not until the Institution of Mechanical Engineers published a comprehensive paper on chrome vanadium steels in 1904 that these alloys received any serious consideration.[34] Even then there was little immediate prospect of its widespread use since its supply was limited and its cost high. By 1905, several types of vanadium steels were being produced in Britain mainly for use in the motor industry, but in this same year deposits were discovered in Peru and later a survey team found at Minas Ragra, high in the Andes, what eventually proved to be the world's largest single vanadium ore deposit. These finds were of the utmost importance for the motor industry. J. Kent Smith, the British metallurgist who did much to promote the use of vanadium steels—and who was associated with the firm of Messrs George Blackwell, Sons & Co., mentioned earlier (p. 6)—visited Henry Ford together with two of his consultants, Harold Wills and Charles Sorenson, in 1906 when he was able to demonstrate to them that the vanadium steel alloy had a tensile strength about three times that of the steel they were using. Furthermore, its toughness was superior and it could be machined more easily than plain carbon steel. So convincing was this demonstration that it induced Henry Ford to begin experimental work which resulted in the Model T, for the construction of which twenty different steels were specified, ten of them being vanadium steels. To appreciate the significance of this, it must be recalled that most of the cars manufactured at that time used no more than about half a dozen different types of steel. Vanadium steels were used for crankshaft, camshaft, connecting rods, differential spider, transmission shafts, axles, gears, valve springs and many other components.

Not only was this versatile alloy an important new material in the fabric of the car, it was a feature emphasized in the advertisement of the vehicle as the following extract from the Ford Sales Catalogue for 1909 shows:

Vanadium Steel

The Model T is built entirely of the best materials obtainable. No car at $5000 [the Model T then cost $850] has higher grade, for none better can be bought. Heat treated Ford Vanadium through-

> out; in axles, shafts, springs, gears—in fact a Vanadium steel car—is
> one evidence of superiority.
>
> Nobody disputes that Vanadium steel is the finest automobile steel
> obtainable. Ask any disinterested steel expert about Vanadium steel
> and listen to him enthuse about its dynamic strength and power to
> withstand sudden shocks, torsional strain and vibration; its tensile
> strength, elasticity, ductility and withal the ease with which it stands
> machining. We defy any man to break a Ford Vanadium steel shaft
> or spring or axle with any test less than 50% more rigid than would
> be required to put any other steel in the junk pile, and that's a
> conservative statement.[35]

Quite obviously tastes in sales literature have changed as much as the
vehicles themselves! The superb qualities of this alloy so euphorically
expounded here were not apparent to American steel-makers and Ford
encountered absolute reluctance to produce it on their part. Only after
Ford had guaranteed to indemnify them against any losses did the United
Steel Company of Canton, Ohio, undertake the task. Without question,
the material has proved its worth.

To meet the requirements of advancing designs, vanadium alloys have
been continuously developed and improved throughout the years. During
the past thirty years, the growing tendency to produce body sections with
complicated contours by pressing has necessitated the production of steels
able to withstand large amounts of deformation. While this may not
appear a particularly difficult exercise, the metallurgical problems involved have been not insignificant. To be amenable to forming on existing
presses, such steels have had to possess an initial strength similar to plain
carbon steel and have a high ratio of work hardening combined with
minimum springback. Freedom from unsightly flow lines on the surface of
the metal after forming has required an alloy with a continuous stress-
strain relationship, that is, without a yield point. Once more a suitable
material has had to be designed for a specific product and again, in this
instance, the requirement has been satisfied by a vanadium steel.

Tool Steels—Further Progress

In 1876, Henri Brustlein of the French Holtzer Steel Works began experiments on the addition of chromium to steels and, having established
its value, it seems that small amounts were added to the original composition of Mushet's self-hardening steel. At the same time, in the USA, Dr
J.W. Langley pointed out, in a paper presented before the American
Society of Civil Engineers, that the addition of tungsten alone to steel did
not endow it with the property of self-hardening and that to acquire this
characteristic the alloy had to contain a high percentage of manganese or
chromium. By the early 1890s, some manufacturers of self-hardening steels
were increasing the chromium content at the expense of manganese and
in the prodigious experimentation which was characteristic of that period

the carbon content of the Mushet steels was lowered significantly. This was the first step in the development of high-speed steels, but due to the secrecy surrounding the activities of tool-steelmakers it has never really been established who was originally responsible for suggesting this reduction. The development of high-speed tool steel is generally attributed to F.W. Taylor and M. White, although according to a recent writer 'Taylor never claimed to be the originator of the steel, and at the time he recommended high-speed steels of lower carbon content, they had been made in England for at least three to five years'.[36]

Certainly, however, an astonishing exhibit at the Paris Exhibition of 1900 was a tool of one of these alloys cutting with its nose at a red heat, an impressive sight which has been repeated in numerous exhibitions ever since. Hundreds of different high-speed steels have been developed and used, all sharing the characteristic of great hardness combined with a resistance to the tempering effect of the heat generated during cutting, a feature which has permitted cutting speeds far greater than those possible with ordinary carbon tool steels. These properties are achieved by the inclusion of the refractory metals, chromium, cobalt, molybdenum, tungsten and vanadium, all of which form very hard and stable complex carbides. Indeed, many of the varieties produced have eliminated ferrite from their compositions and are thus not 'steels' at all. The cemented carbide tools, often consisting of a cobalt matrix with complex tungsten-chromium carbides, are formed from metal powders by sintering at high temperature and pressure in an atmosphere of hydrogen, to give a structure which resembles that of a high-speed steel.

Success in producing high-speed steels at the turn of the century, while solving a problem, created another of equal magnitude. Although capable of cutting at dramatically higher speeds than the commonly-used alloys, they could not do so on the machines available because the machines were unable to withstand the stresses and vibrations induced. And so the new tool alloys triggered innovations in the design of machine tools which, in general, became much heavier, faster, and more rigidly constructed, a trend which besides extending the range of their use facilitated their introduction to new uses. Better tool alloys and the growing requirements of motor-car production provided, in combination, an extraordinary stimulus to innovation and improvement in the whole range of machine tools.

Summary

Steel alloys only really became a commercial proposition after steel of consistent quality and in great quantity had finally achieved a pre-eminence over wrought iron, and this did not occur until the final decade of the nineteenth century. Tungsten tool steels had established themselves prior to this, of course, but they were only of small quantity production and, indeed, remain so today when cutting-tool alloys still represent less than 1 per cent of alloy steel production.

It was naval armament which first showed the advantages of alloy steel for large structural purposes. The desire for military superiority and the demands of warfare have been a constant stimulus to the search for and production of alloy steels throughout the past century. The establishment of alloy steel industries in the European nations with, in many cases, a dependence on the importation of the necessary alloying elements, gave immense political and strategic significance to remote parts of the world. Countries rich in mineral deposits such as the Belgian Congo (cobalt), Bolivia (tin), Burma (tungsten and tin), Canada (nickel), the Malay Peninsula (tin), New Caledonia (chromium, cobalt, nickel, manganese, platinum), Peru (vanadium) and Rhodesia (chromium), became very important as sources of raw materials vital to the economies and industries of the major industrial powers. Possession of minerals has, itself, been a contributory factor to the prosecution of war. The possession of Lorraine, whose deposits of phosphoric iron ores acquired an enhanced value with the introduction of the basic Bessemer process, has been an issue of contention in every conflict between France and Germany. Similarly, British diplomatic success in purchasing Yugoslavia's whole output of chrome, early in the Second World War, helped to precipitate Hitler's invasion of the Balkans in 1941, while the need for manganese, of which the Caucasus had rich deposits, led Germany into what proved to be a disastrous campaign.

Warfare has brought about the development of new alloys because of the shortage of alloying elements on the one hand and the intensive effort to produce better and more powerful weapons, or the means of using them, on the other. When the Pacific was overrun by the Japanese and Britain's major sources of tungsten were lost, the shortage was alleviated by the partial substitution of molybdenum for tungsten in high-speed steels. Molybdenum, like tungsten, promotes the formation of very hard carbides upon which the success of the alloys depend, but the molybdenum alloys, being more susceptible to decarburization during heat treatment than the tungsten alloys, require a more closely controlled heat treatment which is now achieved with modern plant. Since the molybdenum high-speed steels are considerably tougher than the corresponding tungsten types and are now widely used, it may be argued that the temporary deprivation of the usual alloying element at least accelerated the advance to a better product. However, not all avenues of exploration led to success. Rumours during the First World War that the Germans had developed zirconium steels with wonderful properties for ordnance instigated research on the part of several governments and private companies into the reduction of the element and the properties of zirconium alloys, but the results were not particularly encouraging. Zirconium could not be reduced successfully at that time and it was some thirty years later before work, under the direction of Dr W.J. Kroll in the USA, produced a viable commercial process similar in some respects to the method employed for the production of titanium. The solution to the problem involved the development of vacuum techniques which were subsequently applied to improve titanium production.

So, too, H. Brearley's attempt to improve the erosion-resisting properties of steel employed for rifle barrels led to the discovery of stainless steel which, while not being the answer to his immediate problem, was to acquire great importance within a few years. Indeed, the 1920s were a very important period in the history of steel alloys. Early in the decade the stainless and high chromium nickel steels first came into commercial production and then, with the introduction of heat-resisting steels, the struggle to attain increased strength and corrosion resistance at ever-higher operating temperatures began and has continued with mounting intensity ever since.[37] Besides prime movers in their various forms and manifold applications, calling for specially-designed and produced materials, advancing electrical technology and particularly wireless communication and public broadcasting services gave great impetus to the development of alloys with specific electrical and magnetic characteristics.

Through equipment such as the electric arc furnace, the high-frequency induction furnace and the vacuum furnace, together with heat-treatment processes such as precipitation-hardening and nitriding (discovered in 1923), the production of high-quality steels amenable to specific and closely-controlled finishing operations to impart or enhance particular qualities has become a practical proposition.

The planned development and exploitation of alloy steels is almost entirely an achievement of the twentieth century, and during this period the third decade is of crucial importance for then some earlier work came to fruition while the seeds were sown for future achievement. While in some specialized applications, notably those involving heat-resisting properties, magnetic performance and cutting-tool materials, the peak operating requirements can no longer be met by steel alloys, there is no doubt that for innumerable uses alloy steels will have a major and vital role to play in the future.[38] Many new ones are being, and will continue to be, developed to meet ever more-exacting performance specifications, manufacturing processes, heat-treatment requirements and economic limitations. As the cost of alloying elements goes up and the world's resources diminish, it seems imperative to ensure the maximum efficiency in the use of materials, with the possession of unnecessary properties as unacceptable as the lack of essential characteristics.

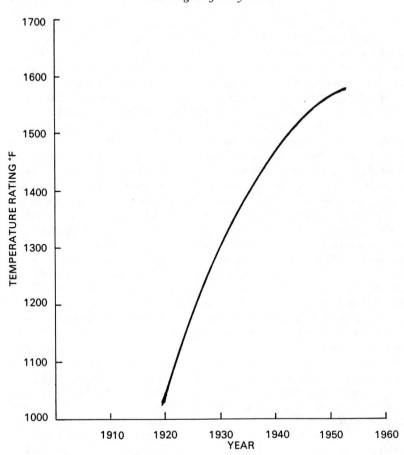

Graph 2. Stress-rupture properties of steels compared with other alloy systems
(1,000-hour rupture value).

Appendix 2

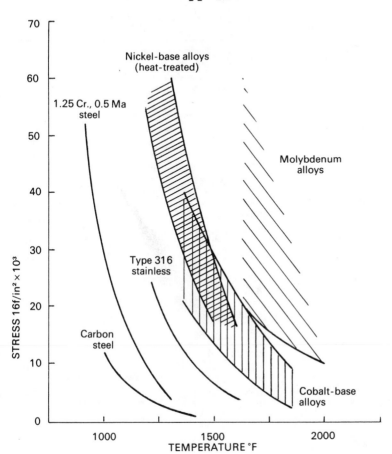

Graph 1. Chronological increase in temperature ratings (100-hour life at 20,000 lbf/in²) of turbosupercharger and jet-engine bucket alloys.

The Origins of Alloy Steels

Appendix 1

(All figures are taken from *Abstract of British Historical Abstracts* by B.R. Mitchell and Phyllis Deane, Cambridge University Press, 1962)

Textiles	Spindles, Power looms, and Power (in thousands)			Horse power used	
	Year	Spindles	Power looms	Steam	Water
Cotton factories (Table 6, p. 185)	1835	—	110	—	—
	1838	—	—	46	12
	1850	20,977	250	71	11
	1856	28,010	299	87	9
	1861	30,387	400	281	12
Wool factories (Table 16, p.198)	1835	—	5	—	—
	1838	—	—	17	10
	1850	2,471	42	23	10
	1856	3,112	53	31	9
	1861	3,472	65	53	11
Flax, jute, hemp and china grass factories (Table 21, p. 203)	1835	—	0·3	—	—
	1838	—	—	7	4
	1850	965	4	11	3
	1856	1,288	8	14	4
	1861	1,252	15	32	4
Silk factories (Table 27, p. 210)	1835	—	2	—	—
	1838	—	—	2	1
	1850	1,226	6	3	1
	1856	1,094	9	4	1
	1861	1,339	11	6	1

Railways (Table 5, p. 225)	Year	Miles of track
	1825	$26\frac{3}{4}$
	1830	$97\frac{1}{2}$
	1835	$337\frac{3}{4}$
	1838	$742\frac{1}{2}$
	1840	$1,497\frac{3}{4}$
	1850	6,084
	1856	7,650
	1861	9,446

Notes and References

1. Vanadium was actually discovered by the Spanish-Mexican mineralogist A.M. Del Rio, in 1801, and given the name erythronium. However, since other chemists considered erythronium to be chromium (discovered in 1797 by L.N. Vauquelin), Del Rio was influenced by their opinions and did not maintain his claim for the discovery of the new metal, which is usually credited to Sefström.

2. The figures given in Appendix 1 provide some measure of the rapid rate of mechanization in the United Kingdom towards the middle of the nineteenth century.

3. J. Dearden, *Iron and Steel Today*, Oxford UP, 1956, p. 190.

4. C. Singer, et al., *A History of Technology*, Vol. 5, Oxford UP, 1970, p. 65.

5. Sir R.A. Hadfield, *Faraday and His Metallurgical Researches*, Chapman & Hall, London, 1931, p. 260.

6. The metals and non-metallic substances used by Faraday in his experiments are quoted by Sir Robert Hadfield in his book, note 5, p. 108, as follows:

Metallic elements:

1. Chromium
2. Copper
3. Gold
4. Iridium
5. Iron
6. Nickel
7. Osmium
8. Palladium
9. Platinum
10. Rhodium
11. Silver
12. Tin
13. Titanium

Non-metallic elements:

14. Carbon
15. Silicon
16. Sulphur

Other ferrous substances:

17. Wootz or Indian steel
18. Carburet of iron
19. Alumine alloy, so termed by Faraday in 1820, in imitation of wootz, also described as damascene steel (Davy spoke of the added element as aluminium)
20. Meteoric iron.

7. This assessment from the paper 'On the Alloys of Steel' by Faraday and Stodart (1822) is quoted by Sir Robert Hadfield, note 5, p. 118.

8. J.C. Carr and W. Taplin, *History of the British Steel Industry*, Basil Blackwell, Oxford, 1962, p. 20.

9. Sir R.A. Hadfield, as note 5, p. 261.

10. *Ibid.*, p. 262.

11. J.C. Carr and W. Taplin, as note 8, p. 222.

12. C.S. Smith (ed.), *The Sorby Centennial Symposium on the History of Metallurgy*, Gordon & Breach Science Publishers, London, 1965, p. 184.

13. *Ibid.*, p. 184.

14. J. Dearden, as note 3, p. 29.

15. C.S. Smith (ed.), as note 12, p. 469.

16. H.C. Sorby, 'On the Application of Very High Powers to the Study of the Microscopical Structure of Steel' in *Journal of the Iron and Steel Institute*, Vol. XXXI, 1886, pp. 140-4.

17. C.S. Smith (ed.), as note 12, pp. 183-4.

18. J.M. Camp and C.B. Francis (eds), *The Making, Shaping and Treating of Steel*, 6th edn, United States Steel Company, Pittsburgh, 1951, p. 1290.

19. Sir R.A. Hadfield, *Metallurgy and its Influence on Modern Progress*, Chapman and Hall, London, 1925, pp. 113–16.

20. J.M. Camp and C.B. Francis (eds), as note 18, p. 1290.

21. J. Dearden, as note 3, p. 223.

22. *Ibid.*, p. 230.

23. Sir R.A. Hadfield, as note 5, p. 264.

24. C.S. Smith (ed.), as note 12, p. 84.

25. Sir R.A. Hadfield, as note 5, p. 266.

26. *Ibid.*, p. 272.

27. J. Hopkinson, 'Magnetic Properties of Alloys on Nickel and Iron' in *Proceedings of the Royal Society*, Vol. 47, 1899.

28. R.A. Higgins, *Engineering Metallurgy, Part 1*, English UP, London, 1968, p. 292. Table 14.2 quotes the following values for BHmax (mega-gauss-oersted): 1 per cent carbon steel (hardened, 0·2; 'Columax' (an anisotropic alloy), 7·5.

29. H.W. Dickinson, *A Short History of the Steam Engine*, Cambridge UP, 1938, p. 226. The figures quoted here are taken from the graph on p. 226. The temperatures quoted are those at the throttle.

30. A.T.V., *Acier Turbines à Vapeur*.

31. Sir R.A. Hadfield, as note 5, pp. 290–1.

32. *Ibid.*, pp. 293–4.

33. David F. Nicholas, *Vanadium steel and the car that changed the world*, Foote Prints, Vol. 39, No. 1, 1973, Foote Mineral Company, Exton, Pa., pp. 2–15.

34. Between 1899 and 1902, Professor J.O. Arnold, in Sheffield, carried out experiments which showed that as little as 0·15 per cent vanadium added to nickel steels, chrome steels and nickel chrome steels effected significant increases in tensile strength (J.C. Carr and W. Taplin, *op. cit.*, p. 222).

35. David F. Nicholas, as note 33, p. 8.

36. G.A. Roberts, J.C. Hamaker and A.R. Johnson, *Tool Steels*, American Society for Metals, 1962, pp. 6–7.

37. The application of superchargers to aircraft engines, commencing in 1916, marks the beginning of the real quest for high-temperature alloys and Graph 1, in Appendix 2, shows the increasing temperature rating of alloys used in aircraft engines. The levelling off of this curve around 1950 is due primarily to the ductility requirements of turbine blades preventing further hardening additions to nickel- and cobalt-base alloys. This diagram is based on a graph of temperature ratings given in R.F. Hehemann and G. Mervin Ault (eds), *High Temperature Materials*, John Wiley & Sons, Inc., New York, 1959, p. 408.

38. An indication of the relative stress-rupture properties of steels and other high-temperature alloy systems is given in Appendix 2, Graph 2. Source: *High Temperature Materials*, as note 37, p. 8.

Bibliography

J.C. Carr and W. Taplin, *History of the British Steel Industry*, Basil Blackwell, Oxford, 1962.

J. Dearden, *Iron and Steel Today*, Oxford UP, 1956.

T.K. Derry and T.I. Williams, *A Short History of Technology*, Oxford UP, 1960.

Sir R.A. Hadfield, *Metallurgy and its Influence on Modern Progress*, Chapman and Hall, London, 1925.

Sir R.A. Hadfield, *Faraday and His Metallurgical Researches*, Chapman and Hall, London, 1931.

R.A. Higgins, *Engineering Metallurgy, Part I*, The English UP, 1968.

B.S. Keeling and A.E.G. Wright, *The Development of the Modern British Steel Industry*, Longmans, 1964.

M. Kranzberg and C.W. Pursell (eds), *Technology in Western Civilization*, Vol. I & II, Oxford UP, New York, 1967.

B.R. Mitchell and Phyllis Deane, *Abstract of British Historical Statistics*, Cambridge UP, 1962.

D.F. Nicholas, *Vanadium steel and the car that changed the world*, Foote Mineral Company, Exton, Pa., 1973.

E.C. Rollason, *Metallurgy for Engineers*, Edward Arnold, London, 1961.

C. Singer, E.J. Holmyard, A.R. Hall and T.I. Williams (eds), *A History of Technology Vol. 5*, Oxford UP, 1958; *Vol. 6*, Oxford UP, 1978.

A Fourteenth-Century Pile Driver: the *Engin* of the Bridge at Orleans*

MARJORIE NICE BOYER

Medieval pile drivers have not been a subject which has excited much interest either in their own day or in ours. Consequently, there is a paucity of evidence on the subject. The first pictures of pile drivers come from the fifteenth century and are to be found only in technological treatises, while one must wait for the sixteenth to find illustrations of them in actual use. Most of the written records consist of brief references to pile drivers. However, for the years 1387–9 there have fortunately survived extensive accounts of three pile-driving seasons at Orléans in France. Manuscript Archives du Loiret CC 920, the receipts and expenditures of the corporation of the bridge at Orléans and of the hospital situated on it, contain the cost of purchasing the parts of the pile driver as well as the piles, of repairing the *engin*, and of paying the men who operated it. These accounts are a rarity for our topic. Nowhere else so early do we have so many details on a pile driver at a single site. It is the purpose of the present article to determine, on the basis of the aforementioned evidence, how a fourteenth-century pile driver worked and to compare its operation with those of later periods.

Just how unusual the Orléans records are is made clear by a comparison with the information to be gleaned from other sources. Very little more is known about the pile drivers used at Bloemkamp, in the Netherlands, in 1238 than that there was a four-man crew for each, or about the pile drivers employed at London in the thirteenth and fourteenth centuries than that the accounts mention running rams and gibbet rams.[1] It will come as no surprise to historians of technology to learn that the standard Old French dictionaries, Godefroy and Tobler-Lommatsch, are not very helpful for explanations of medieval pile drivers.[2] More can be learned from contemporary records than from the dictionaries. At Lyon in 1416, the pile driver was called the 'engin à planter des paux'—engine to drive piles—or merely the *engin* of the bridge over the Rhône.[3] What seems to have been common to *engins*, whether an *engin de guerre* or an *engin* to raise a portcullis or one to drive piles, were the pulley and rope.[4] At Albi[5]

* My thanks for criticism, suggestions and information go to Norman A.F. Smith, T.H. Boyer, Dorothy De Flandre, Lewis J. Bodi, Andrea Matthies and Bert S. Hall.

A version of this paper was read before the Washington meeting of the Society for the History of Technology on 21 October 1983.

(1408-10) it was the rope that was the essential part of the pile driver. Some information on the parts of the *engin* of the bridge over the Rhône is furnished by the records of the town council of Lyon. In 1421 a flood had carried the town's pile driver downstream 22 kilometres to a point where it was gathered in and dismantled. The man responsible for salvaging it was ordered to return 'two beams, an *engin*, and a tower for driving piles'.[6]

In contrast to these fragments on pile drivers is the wealth of detail furnished by the Orléans bridge accounts for 1367-89. The records of the *pont d'Orléans* must be understood in the light of the peculiarities of the site and the architecture of the bridge. The Loire at Orléans is an old river, geologically speaking, and before the construction of levées and dikes was extremely variable in volume, in flood time stretching from bank to bank and in summer sometimes dwindling to small streams in an expanse of sand. In the summer, navigation of the Loire at Orléans ceased. On the island crossed by the bridge stood the hospital and chapel of Saint-Anthoine, both of which were always referred to as 'on the bridge'.

The bridge at Orléans was a twelfth-century bridge and, like its fellows on the Loire at Blois and Tours and like London and Rochester bridges, it was constructed on starlings. This method consisted of driving piles half their length into the river bed to form a crib, dumping in rubble to a height above water level, laying down sleepers, and building masonry piers on this foundation. Constructing stone piers on starlings meant that it was of the first importance to maintain adequate piles to prevent the rubble from shifting and causing the bridge to fall. Hence it was that at Orléans the pile driver was indispensable. Almost every summer a campaign was undertaken, sometimes longer, sometimes shorter, according to the condition of the bridge. At Orléans in 1387-9, the pile-driving season coincided with the period of low water, that is, from late June into the third week in September, although once the *engin* worked two days in October. In a river subject to sudden freshets, even working the first week in September could be hazardous. The fact that the *engin* of the bridge of Orléans operated from a boat involved much concern about floods.

There were three typical medieval pile drivers: the hand *ram*, the *hye* and the *engin*. The hand ram is attested by pictorial evidence alone,[7] while documents of the period confirm the use of the other two. The *hye* and the *engin* each had a frame from which hung a pulley with a rope holding up the ram, but the *hye* had a windlass, while the *engin* did not. The latter's rope ended in a series of individual cords for a crew of men to raise the ram. There are not many examples of the *hye* being employed in medieval France. One occurred during the construction of a timber bridge across the Seine, the Pont Notre-Dame, when the ceremonial equivalent of laying the first stone was performed on the *hye*. On 31 May 1413, King Charles VI struck the pile with a blow of the *hye*, and he was followed by his eldest son, the duke of Guienne, by the duke of Berry, the duke of Burgundy, and by the sire de la Trémoille.[8]

At Orléans an *hye* was built by a wheelwright and was operated one day in March 1389 by six men but then not again for four months. By July, the *hye* had been improved so that it required only two men, but it was apparently still not satisfactory. It was used for two days and then not again.[9] The Orléans *hye*, based on land, drove piles to reinforce the banks of the Portereau at the southern end of the bridge, where the corporation owned land. In contrast, the pile driving for the bridge itself was done by an *engin* placed in a boat. The *engin* was used season after season: for example, in 1387 for 37 days, in 1388 for 26 days, and in 1389 for 57 days. The *hye* was less expensive to operate and required fewer men, but clearly it could not compete with the *engin*.

One can speculate as to the reasons why the *hye* was at a disadvantage. Perhaps it was partly a question of technical expertise. Pile drivers were made on the spot by local craftsmen, both at Orléans and at Poitiers. At the latter town in 1386, when it was a question of repairing a drawbridge, four cartloads of timber, each drawn by four oxen, were transported to Poitiers to make the frame for the ram as well as the piles and other things for the drawbridge of one of the town gates.[10] It may be that local artisans neither knew nor cared to learn the specialized techniques for producing an efficient pile driver with a windlass. There were other reasons for its unpopularity, not only in the fourteenth and fifteenth centuries, but in the seventeenth and eighteenth. Later writers note that it was slower than the *engin*, and while the *hye* managed larger piles and used fewer men,[11] apparently builders preferred to drive piles with a series of lighter blows rather than with fewer, heavier ones. Operators of pile drivers seem to have avoided great piles, and they were indifferent to the idea of saving manpower, a commodity in plentiful supply and relatively cheap.

Engins can be subdivided according to whether the pulley was suspended from the top of an upright or from an arm protruding from it. In the latter case it was a gibbet ram, as shown in Fig. 1, where a French illumination of 1519 shows Caesar directing the building of his bridge across the Rhine. In a gibbet ram, the pile driver head was necessarily free-falling, but where the pulley hung either from the top of the upright or from a point where several uprights met or were connected, the ram might either be free-falling or a running ram. The former is illustrated in *The Anonymous of the Hussite Wars*, folio 37 verso,[12] and an example of the running ram is given in Fig. 2. The picture shows the building of a bridge across the Scheldt in 1584–5 to prevent the Dutch from provisioning Antwerp. In this case the ram is fitted with protruding horizontal rods to keep it in place as it slides up and down between the vertical beams which guide it. It will be noticed that the pile driver is placed in a boat, and yet the ram is a running ram. This is effected by an indentation in the bow to allow for guide rails for the ram, which nevertheless is directly over the pile to be driven.

At Orléans, according to the accounts of 1387–9, the first order of business was to purchase the piles, the second to buy a boat to put the pile driver on, and the third to put the *engin* in working order. Both the piles and the *engin* were of elm, and this wood was used at Westminster for piles in 1338 and at Rochester Bridge in the fourteenth century.[13] Chestnut was preferred at Lyon and Agen and alder at London Bridge.[14] Elm has good resistance to water, but it is probable that the paramount consideration at Orléans in the choice of wood for piles was availability. In an age of high transportation costs, medieval bridges were commonly built of whatever materials were at hand.

In successive years, piles for the bridge at Orléans were purchased from local suppliers in the following quantities:

1387	1388	1389
100 @ 9 ft (2·98 m)	100 @ 10 ft (3·3 m)	50 @ 11 ft (3·63 m)
100 @ 10 ft (3·3 m)	75 @ 11 ft (3·63 m)	100 @ 12 ft (3·96 m)
100 @ 11 ft (3·63 m)	25 @ 13 ft (4·29 m)	25 @ 14 ft (4·63 m)
100 @ 12 ft (3·96 m)		50 @ 15 ft (4·95 m)

In every case the diameter was a span (0·2295 m) at the crown. It may be that the piles varied in length because the timber available in the forest did so, or perhaps different spots required longer or shorter piles according to the depth of the water in that part of the river. When in 1389 it was a question of six great pieces of squared timber $3\frac{1}{2}$, 4 and 5 *toises* long, i.e. 6·828, 7·796 and 9·745 m in length, the only place such long piles could be driven was beside the hospital on the island crossed by the bridge.[15] Perhaps the water was deeper there, and in that case the boat would ride sufficiently high so that the frame could lift the ram to drive such long piles. The size of the piles was very much that recommended later by Perronet in the eighteenth century. He states that 9 *pouces* (0·2295 m) is adequate for the diameter of a pile 9 to 12 ft long (2·97 to 3·96 m).[16] In the fourteenth century, piles were sharpened at Orléans but were not iron shod.

When the piles had been ordered, the next item on the agenda of the bridge wardens was the purchase of a boat to put the pile driver in. They economized. Their object was the purchase of a boat in just good enough condition to last the season and be taken apart in September or October so that its timbers could be laid on the starlings. In 1387 the boat was a *barch* carrying sixteen men and the pile driver, but in 1388 and 1389 it was a *chaland* holding fourteen men and the *engin*. All we know about the size of the *barch* is that it required nine men to lift the boat with the pile driver out of the water when a flood threatened during the week of 2 September 1387.[17] The *chaland* was the large freight boat used on the Loire. A triumph of adaptation to its environment, the *chaland* was a flat-bottomed boat of shallow draught, so that it required very little water for its operation. This was a necessity on the Loire. The size of the *chaland* may be judged by the length of the beams placed on either side to protect it—5·87 m. Perhaps one can surmise that the *chaland* was smaller than the *barch*, because it carried fewer men: fourteen, not sixteen. We know of no other reason for the reduction in personnel: the ram was the same, and most of the piles, except for a few 66 cm longer, were identical in length.

Once the boat had been placed 'under the bridge', the next thing was to assemble the pile driver and put it in the boat. To judge from the items purchased for the *engin*, it was uncomplicated. The pile driver resembled the one in Fig. 1. It must have been a gibbet ram. Such a pile driver required no alterations to the boat, and the expense accounts indicate that the sole cost was the purchase price. Indeed, the slender resources of the corporation of the bridge and hospital of Orléans did not permit of the building and operation of the state-of-the-art pile driver like the one produced in 1584–5 by an expert with funding from the king of Spain, as illustrated here in Fig. 2. At Orléans it may be assumed that the ram was free-falling, but this probably did not involve extreme inconvenience. They may have lifted the ram only a short distance, as is common twentieth-century practice.

At Orléans, further preparation for the opening of the pile-driving season involved fetching the parts of the disassembled *engin* from the chapel of Saint-Anthoine, where they had been stored over the winter. Carpenters put the *engin* in working order, but blacksmiths were necessary to do the iron work. This consisted of iron hooks, bolts, bars and nails. One year at the beginning of the season the wardens bought a piece of elm to make an upright for the pile driver. The timber was carried to a blacksmith so he could reforge the iron and put on two hooks as well as make the *engin's* iron bolt longer. The purpose of the hooks and of the iron bars was to connect the various parts, so we are told. When the parts were ready, they were carried to the boat and bolted together.

The chief material for the pile driver was wood. The frame with its braces, the pulleys and the ram were all of elm. The pulleys frequently had to be replaced. Two were purchased from a turner in the week of 12 August, another two in that of 2 September 1387, two in June 1388 and four in June 1389. Since there were always two pulleys, it is probable that one hung from one end of the extended arm and the other from the junction of the arm and the upright (see Fig. 1). It is to be assumed that the choice of wood for the pulleys at this period was deliberate, for the Orléanais both before and after used metal ones—in 1358 an iron pulley and in 1419 a brass one, the latter on a war engine.[18]

The ram was the indispensable part of the pile driver, so much so that its name was sometimes substituted for the entire machine. Rams were made of wood at Lyon and elsewhere, at London and Orléans of elm wood.[19] In 1387 the wardens purchased an elm stump and had it made into a ram by a carpenter. This *souche*, defined as that part of the tree remaining in the ground after it had been felled, would have been the hardest part of the tree. This ram lasted for two seasons, that is to say 64 days in 1387 and 1388, during which time it drove up to 600 piles. In 1389, a great elm trunk was made into two rams. At Orléans there is no evidence of applying iron to the ram, although in the next century the *engin* in *An Anonymous of the Hussite Wars* on folio 37 verso has a ram which appears to be iron-bound.

In addition to the wood for the ram, frame and pulleys, and iron for

Figure 1. A gibbet ram in a boat. Caesar building a bridge across the Rhine. British Museum, Harl. 6205, Folio 21, from the sixteenth century. (By permission of the British Library.)

hooks, bolts and other things, purchases were made of leather straps to hang the ram, ropes and cords to raise it, and lubricant for the *engin*.[20] There was a rope to moor the boat, small cords, one for each man, and a number of great ropes to pull the *mouton* (ram). In the campaign of 1387, four great ropes were purchased for this purpose, in 1388 three, and in 1389 six. It would seem that the great ropes used for the ram did not last long and that the work was very hard on them. Quantities of lubricant for the *engin* were bought, apparently to apply to wooden pulleys and the ropes, the only moving parts of the pile driver other than the ram. In 1387, $31\frac{1}{2}$ lb were used in a season of 37 days; in 1388, 27 lb in 26 days; and in 1389, $44\frac{1}{2}$ in 57 days; an average of 1 lb 2 oz or 0·53 kilograms of lubricant daily. At the end of the season the pile driver was 'descheville et desoint', that is, unbolted and the lubricant removed, before being stored in the chapel of Saint-Anthoine.

To keep the pile driver running required constant attention. Part of this was the application of pounds of lubricant and the replacing of ropes and pulleys, but there were also frequent notations that the pile driver was *despecie*, coming to pieces, and that it had to be repaired. Often the iron work had to be redone and the parts of the frame replaced. Carpenters seem to have been kept busy counteracting the tendency of the *engin* to shake itself to pieces and to try to keep the pile driver running continuously all season.

Operations were under the direction of one of the bridge wardens, whose function it was continuously to supervise the workmen. The first point was to assemble them in the boat under the bridge, for the pile driver was chiefly employed around the starlings, although sometimes to reinforce the banks of the island in the middle of the bridge or of the Portereau, the area at the southern end of the bridge. The piles were stored either on the bridge or at the Portereau and carried onto the bridge. From it, two men let them down with a rope, one by one. They were received, upended and held in place by two men in a small boat, for it was their job to 'arrange the piles'. When the ram was directly over the pile, the bridge warden gave the word, the fourteen or sixteen men pulled on their ropes to raise the ram and, when the signal was given, suddenly let go to allow it to strike the pile. Figure 1 shows the big boat with the pile driver and the men, each with his cord. Standing in a small boat, a man is holding the pile to be driven, and two other men have more piles ready for the pile driver. At Orléans, when a pile had been driven, the boat had to be untied, moved to a new place and again moored to the starling. Then the process of pile driving recommenced. Finally, a wheelwright and his varlet bolted the piles to boards so as to hold them in line.

At Orléans the sandy bottom should have been easily penetrated by piles merely sharpened at the end rather than iron shod. Yet the number of men (fourteen to sixteen) was greater than the four shown in Fig. 1, a French illumination of 1519, or the eight pulling on the ram in the pile-driver boat at Antwerp (1584–5) in Fig. 2. Later writers consider that anywhere from twelve to thirty men are necessary to operate an *engin*.[21] In 1782–3, Perronet wrote that twenty-four to twenty-eight men can pull a 600 to 700 lb ram.[22] If the latter means that twenty-four men pulled 600 lb and twenty-eight men 700, then one man pulled 25 lb (111 Newtons corresponding to lifting a mass of 11·4 kg). Then at Orléans, sixteen men could have pulled a ram weighing 400 lb and fourteen men one weighing

Figure 2.   Running a ram in a boat. Royal Library, Brussels, MS 7373 (19611), Item 64. Pierre Le Poivre, Recueil de plans de ville et de châteaux. Reigns of Charles V, Philip II and Albert and Isabella, 1585–1622. (Courtesy of the Royal Library, Brussels.)

360 lb. Using a density of elm of 35 lb per cubic foot, and assuming that the ram was a cylinder about 4 ft long, the ram weighing 400 lb would have had a diameter of about 1·9 ft. If we assume this ram was a cylinder with a height equal to its diameter, then the diameter was about 2·4 ft.[23]

It is probable that at Orléans, as elsewhere, a pile was driven with a volley of blows. Such a volley consisted of 20 to 22 blows, according to Lamprecht Gerritsz, an Amsterdam carpenter who in 1595 secured a ten-year patent from the States-General for an improved pile driver need-ing only six to seven men.[24] Each of the six or seven such volleys necessary to drive a pile was followed by a rest pause. Perronet wrote that the men raise the ram $4\frac{1}{2}$ ft (1·485 m) 25 to 30 times a minute, after which they rest an equal length of time. All we can say of the Orléans pile driver is that it drove an average of seven to nine piles daily in three seasons. We do not know if the men took rest pauses or how many volleys of blows were required to drive a pile, but it can safely be assumed that the men who pulled on the ram spent a considerable time waiting around while everything was being readied for the next pile. Pile driving must have been a slow process and perhaps exasperating. It is clear that the bridge warden thought the men were not working as hard as they should. On one occasion he bought them some wine so that they should be more diligent in this matter.[25]

During at least the next 125 years the *engin* continued to be used at Orléans. In 1402–3, a boat was bought for the *engin et mouton* (ram) to drive piles, and during the English siege the *engin* was stored in town in a room rented for it. After the departure of the enemy on 7 May 1429, it was returned with its chest to the bridge.[26] In the middle of the century a special little house was built for the pile driver on the Motte Saint-Anthoine, and in 1513–14 the Orléanais were still using the *engin* to drive piles, this time in the river for a submersible dam. The rams were still of elm, and there were purchases of piles, bolts, nails, chains, cords, ropes and pulleys. Now, however, the *engin* is lubricated both with *oint* and with olive oil, but as far as one can tell from a summary (all that has survived of the accounts),[27] otherwise the operation and parts of the *engin* seem to have been the same as in the fourteenth century.

The Orléanais were highly experienced with the pile driver, essential as it was to the yearly maintenance of their bridge. If they preferred the simpler type, the *engin*, to the more technically sophisticated one with the windlass, the *hye*, it was apparently not that they were averse to the introduction of new ideas as such. When in 1435 the north end of the bridge was ruined by a flood, they founded the new piers, not on the old-fashioned starlings, but with the current method, that is, on piles driven into the river bed.[28] However, although they modernized the piers of their bridge, they did not do the same with their pile drivers. They seem to have found the *engin* effective. They were inured to the difficulties of operating it and reconciled to the circumstance that it required constant repairs. They were convinced that in pile drivers there was nothing like the time-tested *engin*.

Notes

1. Franz Maria Feldhaus, *Die Maschinen im Leben der Völker*, Basel and Stutt-gart, 1954, pp. 197–8. L.F. Salzman, *Building in England down to 1540: a Documentary History*, Oxford, 1952, p. 328.
2. Tobler-Lommatsch, *Altfranzösischen Wörterbuch: Adolf Toblers nach gelassene*

Materialen bearbeitat ... Erhard Lommatsch, Berlin, 1825–71, s.v. Frédéric Gode-froy, *Dictionnaire de l'ancienne française et de tous ses dialectes du IXe au XVe siècle*, Paris, 1890–1902, s.v.

3. Marie Claude Guigue (ed.), *Registres consulaires de la ville de Lyon ou Recueil des délibérations du conseil de la commune de 1416 à 1423*, Lyon, 1882, pp. 7, 565. *Inv. som. arch. com. Lyons*, Vol. 2, p. 189, 1432–35.

4. Philippe Mantellier, 'Mémoire sur la valeur des principales denrées et mar-chandises qui se vendaient ou se consommaient en la ville d'Orléans au cours des XIVe, XVe, XVIe et XVIIIe siècles', in *Mémoires de la Société archéologique et historique de l'Orléanais*, Vol. 5, 1862, p. 412. *Inv. som. arch. com. Loiret. Orléans*, p. 89. Ms. CC 542.

5. Archives communales d'Albi CC 168 folio 72 recto.

6. Marie Claude Guigue, as note 3, pp. 288–9.

7. Attilo Mori and Giuseppe Boffeto, *Piante e veduto di Firenze: studio storico, topografico, cartografico*, Florence, 1926.

8. A. Tuetey (ed.), *Journal d'un bourgeois de Paris*, Paris, 1881, pp. 30–1.

9. Archives du Loiret CC 920 folio 30 recto and 32 verso.

10. Archives municipales de Poitiers Cahier 25 J 7, 9, 14.

11. Cornelis Meijer, *Traité des moyens de rendre les rivières navigables*, Amsterdam, 1696, p. 17.

12. Bert S. Hall, *The Technological Illustrations of the so-called 'Anonymous of the Hussite Wars' in Codex Latinus Monacensis*, Wiesbaden, 1979, folio 37 verso.

13. L. F. Salzman, as note 1, pp. 84–5. M. Janet Becker, *Rochester Bridge, 1387–1856: A History of the Early Years, compiled from the Wardens' Accounts*, London, 1930, p. 7.

14. Marie-Claude Guigue, 'Notre-Dame de Lyon: recherches sur l'origine du pont de la Guillotière et du grand Hôtel-Dieu et sur l'emplacement de l'hôpital fondé à Lyon, au VIe siècle, par le roi Childebert et la reine Ultragothe' in *Mémoires de la Société littéraire de Lyon*, 1874–5, p. 252. O. Fallières, 'Le Pont d'Agen en 1381' in *Congrès archéologique, 68e session*, Agen, 1901, p. 437.

15. Archives du Loiret CC 920 folio 31 verso.

16. Jean Rodolphe Perronet, *Description des projets de la construction des ponts de Neuilly, de Mante, d'Orléans et autres*, Paris, 1782–3, Vol. 1, p. 93.

17. Archives du Loiret CC folio 25 recto.

18. Philippe Mantellier, as note 4, pp. 355, 412.

19. *Inv. som. arch. com. Lyon*, Vol. 2, No. 189.

20. Leather straps bought for the pile driver included traces for ploughs, for carts and one for a pack horse. It is probable that these straps were used to hang the ram, as in England in 1329 a sealskin was used for this purpose. L.F. Salzman, as note 1, p. 328.

21. Frieda van Tyghem, *Op en om de Middeleeuwse Bouwwerf de Gereedschappen en Toestellen Gebruikt bij het bouwen van de Vroege Middeleeuwen tot omstreeks*, Brussels, 1966, Vol. 1, p. 240. Cornelis Meijer, p. 16. Jean Rodolphe Perronet, Vol. 1, p. 97.

22. Jean Rodolphe Perronet, Vol. 1, p. 97.

23. Information and calculations have been supplied by Norman A.F. Smith and T.H. Boyer.

24. G. Doorman, *Octrooien vor uitwindingen in de Nederlandern (de 16ede tot 18de eeuw)*, Den Haag, 1940, p. 93. 1-9-1595. folio 119, G. 21.

25. Archives du Loiret CC 920 folio 22 recto.

26. *Inv. som. arch. com Orléans*, Vol. 1, p. 193. Archives du Loiret CC 930 folios 22 verso, 31 recto, and verso.

27. Philippe Mantellier, pp. 324, 444.

28. Marjorie N. Boyer, 'Moving ahead with the fifteenth century: New Ideas in Bridge Construction at Orleans' in *History of Technology*, Vol. 6, 1981, pp. 1–22.

Rail Stresses, Impact Loading and Steam Locomotive Design

MICHAEL DUFFY

Introduction

This paper examines the in-service working of railway rail by periodic loads caused by the reciprocating and revolving masses featuring in virtually every steam locomotive since 1804.[1] This matter attracted the attention of engineers, and analysts of industrial mechanics, from the beginnings of the steam railway. But although the scientific and quantified investigation which treated locomotive and rail as one system, and which used mathematical methods, had been explored in the work of Clark, Le Chatelier and Lomonossoff[2] it was not fully applied until the final phase of steam traction around 1940. As Schivelbusch relates,[3] the successful union of locomotive and permanent way, first achieved by George Stephenson and his collaborators on the Liverpool and Manchester Railway in the 1830s, resulted in a system termed 'Machine-Ensemble' by the early nineteenth-century analysts of technology. The steam-railway Machine-Ensemble was regarded as a new kind of technology, exemplifying theory and practice in the most advanced degree. The railway Machine-Ensemble served as exemplar until the 1890s, when the methods, organization and equipment of electrotechnology took its place, at least in heavy engineering. This paper traces the history of this machine-ensemble from the successful union of the two major components—the steam locomotive and the track—through the growth of 'systems disharmony' caused by features intrinsic to the locomotive form, to the eventual breakdown of harmony which signalled the obsolescence of steam traction.[4]

This history is instructive, for it emphasizes the great value of close co-operation between academic institutions and industry; quantified investigation; exchange of ideas between different groups whether engaged in engine design, track maintenance or laboratory research. Though not placed on a truly scientific basis until the end of steam traction, these investigations fostered a co-operative approach subsequently employed with other traction systems with track-working problems peculiar to themselves.

A detailed historic review would relate three stories: the evolution of the rolled-steel standard edge rail from the early cast-iron plate and other primitive forms;[5] the development of the steam engine down to the late nineteenth century;[6] and the rise of a theory of rail working. The nineteenth-century contributions of Nollau, Le Chatelier, Clark, Rankine, Reynolds and Dalby need to be summarized, along with the twentieth-century labours of Lomonossoff, Moore, Arnold, Cox, Colam and others, who united the fruits of practical and theoretical investigations into one scientific theory, validated by extensive testing in the laboratory and under operating conditions. The influence of track-working theory on the final generation of steam locomotive designs would then follow. As the evolution of the standard rail forms and permanent way structures have been described in papers published elsewhere;[7] and as there are several detailed accounts of the evolution of the mechanical form of the steam locomotive given by engineers such as Colburn[8] or Chapelon,[9] as well as provided by careful chronologers,[10] this paper will concentrate on the scientific investigation of rail working by locomotives, and the consequences for design.

It is also the purpose of this study to demonstrate a method of investigating the history of technology which affords practically useful lessons for the present-day engineer by deepening his insight into the influences which in the past have shaped design and practice.

Evolution of the Machine-Ensemble

It was known from the beginning of the steam railway that some forms of locomotive engine worked rail, track bed and civil engineering structures worse than others because there were constant derailments due to broken rails and plates, distorted trackwork and damaged bridges on the pioneer systems between 1804 and 1830.[11] These difficulties led to the withdrawal of locomotive traction in certain cases and the use instead of stationary engines which greatly reduced the rail-working and track-bed disturbing forces. But the ultimate success of locomotive traction was assured by the work of the Stephensons and their associates on the Liverpool and Manchester Company during the 1830s, where there emerged a form of permanent way which has become the world standard, together with the archetypal form of steam locomotive. In this phase, engine and track were regarded as a single system, and it was openly declared that there was no sense in studying each in isolation. Though this unified 'overview' was to some extent lost in later decades with the separation of the functions of mechanical and civil engineer, it was always evident in the work of the great railway theoreticians such as Lomonossoff. For the purposes of this paper, the locomotive is treated primarily as a rail-working device.

The complex history of the metal rail can be simplified under five main headings: cast-iron plate; wrought-iron plate; cast-iron edge rail; wrought-iron edge rail; and rolled-steel edge rail.[12] A more detailed account would need further subdivision to include compound rail, bar and strap rail, tramway rail, and rack rail, and each would need referring to contemporary locomotive design and theories of rail loading. Lack of space prevents discussion of any form other than the iron and steel long-length, rolled rails of parallel cross-section, which became the norm after the middle of the nineteenth century. The history of rail form is a complex one, and it must be appreciated that what follows is a simplified, compressed summary only.

The introduction of cast edge rail, made in short sections, is associated with Jessop's 1798 design, which had a varying cross-section, and was laid without a stringer between end-supports resting on blocks of wood, stone or transverse sleepers.[13] Length was approximately $2\frac{1}{2}$ ft (0·761 m) so that joints were very frequent which, with the absence of fishplates, caused non-alignment leading to wheel-shock at the uneven joints, fracture, and derailment on the waggonways. Jessop's rail underwent much improvement in the early nineteenth century, but edge rail was rare before 1810, especially before 1800 when cast plate was more common. Losh and Stephenson produced a modified cast-iron edge rail in 1816,[14] with scarfed joints, secured with a bolt, and chair-supported, though being made in short lengths there were still shock-producing joints at each pair of blocks.

This Stephenson 'fishbelly' or bow-form rail was later produced in longer lengths by rolling, using eccentric rolls to produce the increase in depth and change in section between the supporting points which were distanced as before. This was produced in 15 ft (4·07 m) lengths, weighing 35 lb/yd (17·36 kg/m) and was tested on the Liverpool and Manchester Railway in 1830 in a series of comparative trials of diverse types of rail, fastening and sleeperage, which led to the selection as standard of the rolled rail of constant cross-section, the Locke double-headed section, forerunner of the once-standard British 'bull-head' rail.[15] These longer lengths did much to foster the harmony of the machine-ensemble, and by 1837 the Liverpool and Manchester Railway had been relaid with parallel-section, long-length rail on transverse wooden sleepers.

The parallel-section rolled edge rail developed along two distinct paths (Fig. 1). There was the Stevens–Vignoles, or flat-bottomed rail, dating from the 1830s, which could be spiked direct to the sleepers or stringers: its direct descendant, laid with rail pads, anchors and other modern fittings is the world standard.[16] The United Kingdom standard, sometimes used in Continental Europe and the British Colonies, was the 'bull-head' section, or modified Locke rail, laid with heavy chair, key, and screw or through-bolt fastening.[17] Before the use of modern fastenings and plates with the Stevens–Vignoles rail, the 'bull-head' rail, with chair support, was more resistant to overturning on curves, but this advantage was steadily lost in the 1920s. Although it remained the British standard down to the 1950s, it has—since that date—been gradually replaced by the flat-bottomed form, though much remains to be seen, for instance on the London Transport system. A third major form, the Brunel bridge-section rail, was demonstrated as an inferior section by deflection tests in the mid nineteenth century, and was obsolete by the 1870s for most purposes. In the United States, there was a period of considerable experiment with rails of diverse form, and the Stevens–Vignoles rail was not accepted as standard until the 1870s.[18]

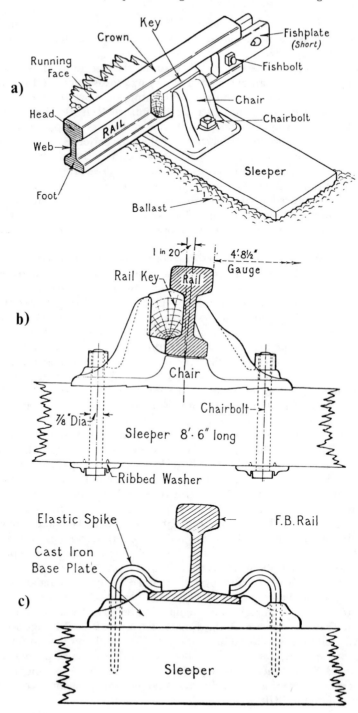

The next major change, spread over several decades, was the substitution of steel for iron, accompanied by a search for better proportions for the basic rail sections.[19] In 1855-6, the Bessemer process made bulk production of cheap steel possible, and railway rails of the new material were experimentally laid in 1857 at Derby and 1861-4 at Crewe and Chalk Farm. The total replacement of iron on main-line systems was a slow business—iron rail on major routes remained in the early twentieth century—though steel was in general use in the United Kingdom by the 1870s. Bessemer steel was produced in the United States by 1865 and was in widespread use by the 1870s, displacing iron from the main lines by the late 1880s.

By 1900, first-class railways had evolved into their modern form of rolled-steel rail, either 'bull-head' or 'flat-bottomed', carried on transverse wooden sleepers without sills, resting on ballast. In the United Kingdom, a typical heavy weight for main-line track subject to severe load was 95 lb/yd (47·1 kg/m) with lengths rolled of 30, 45 and 60 ft (9·15, 13·72 and 18·3 m). In the United States, heavy-weight rail was about 100 lb/yd (49·61 kg/m) for rolled lengths of about 33 ft (10·05 m), and in both countries the average weight of rail in service was considerably less. In 1900, the weight of American locomotives did not exceed that of the British and European by the factor of 3 or 4 that it was to do by 1940, when the US driving axle load was 50 per cent higher than in the United Kingdom, and when a heavy US rail weight was 131 lb/yd (64·92 kg/m).

In the 1970s, the US rail weight was in the range 119–136 lb/yd (59–67·43 kg/m), and most British main-line rail was of a standard 110 lb/yd (54·55 kg/m).

The evolution of the steam locomotive as a rail-working device can be simply classified as follows: plateway locomotives with a single cylinder placed horizontally or vertically; edge-rail locomotives with two cylinders placed vertically; edge-rail locomotives with two cylinders inclined, or horizontal. Once this latter type emerged, on the Liverpool and Manchester Railway, in the form of Robert Stephenson's 'Northumbrian' and

Figure 1. Track structures in common use, 1945. (a) British 'bull-head' rail, chair and fastenings. (b) Cross-section of 'bull-head' rail and associated track structure, showing a type originating on the Great Western Railway (UK) with a serrated chair base to prevent lateral movement and to reduce strain on the chair bolt. Other versions used screws, or spikes, to secure the chair to the sleeper. British Standard 'bull-head' varied in weight (1945) between 60 lb/yd (29·75 kg/m) to 100 lb/yd (49·61 kg/m) in units of 5 lb/yd (2·48 kg/m), the most usual being 95 lb/yd (47·13 kg/m) on main lines. Chairs weighed approx. 46 lb (20·88 kg). (c) Cross-section of 'flat-bottomed' or Stevens–Vignoles rail, laid on base plates secured by elastic spikes. Cast-iron base plates were 37 lb (16·8 kg); steel base plates were 24 lb (10·9 kg). Like the bull-head section shown above, the flat-bottomed rail is normally inclined at 1 in 20 towards the track centre, bringing the rail centre line at right angles to the coned surface of rolling-stock tyres. (Reprinted from H. Greenleaf and G. Tyers, *The Permanent Way*, Winchester Publ., London, 1948.)

'Planet' of 1830, the archetypal structure of steam locomotive, destined to become universal, was achieved. This first phase is associated with the introduction of springing, and the recognition that the action of the reciprocating and rotating masses could severely damage engine, train and track. Due to lack of space in this paper, details of the evolution of the mechanical form of the steam engine before 1830 are left to the works listed in the references.[20]

It would seem, however, that the need to counter rotating masses was appreciated in the first locomotives, for it is generally agreed that Trevithick's plateway engine of 1804 was provided with a weight, placed on the flywheel rim, to balance the revolving mass which could be considered as rotating with the main jack-shaft.

The vertical cylinder locomotives worked the track very badly, especially as it was difficult to spring them adequately due to their tendency—if sprung—to rock about the longitudinal axis.

Placing the cylinders in the inclined position (or better, the horizontal position) diminished this problem, because the disturbing couple then moved towards the horizontal plane, and the reaction of the wheel flanges on the rail opposed it. This contributed to rail wear, but it enabled all axles to be sprung and reduced the violent rocking about the longitudinal axis, which was noticed in the 'Rocket' of 1829, and caused Robert Stephenson to place the cylinders horizontally in that type of engine and the succeeding 'Northumbrian' of 1830. The inclined cylinders, themselves an improvement on those vertically placed, appeared with George Stephenson's 'Twin Sisters' of 1827, 'Lancashire Witch' of 1828, and were used in the 'Rocket' of 1829.[21]

Though introducing a 'nosing' couple, horizontal cylinders did reduce the direct action of steam pressure on the rail through pistons and connecting rods, and the near-horizontal position became standard. Hackworth's 'Sans Pareil' was one of the last examples of the vertical cylinder type.

From the 'Rocket' and 'Northumbrian', the standard form of steam locomotive evolved on the same railway, the Liverpool and Manchester, which fostered the standard track. The 'Planet' of 1830 was the first with the two cylinders placed inside the frames, under the smokebox, as close together as possible to reduce the nosing couple. The locomotive weighed 8 tons, carried on two guiding or pilot wheels, and two driving wheels, resulting in the wheel arrangement known as 2-2-0. Size and weight increased throughout the nineteenth century, so that c. 1900 a typical British express passenger locomotive was as shown in Fig. 2, a 4-4-0 weighing 75 tons total with tender, in running order.[22] Both inside-cylinder (inside connected) and outside-cylinder (outside connected) types were used all over the world, but in the United Kingdom during the nineteenth century there was a preference for inside-cylinder types, whereas in the United States the majority of locomotives were outside cylinder.[23] In the twentieth century the trend in all countries was towards outside-cylinder types for ease of maintenance.

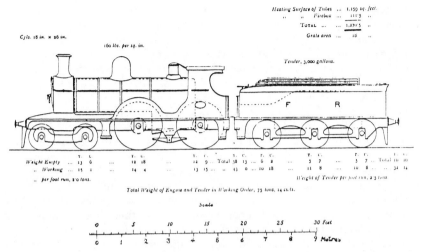

Figure 2. Typical express passenger engine as running in the United Kingdom at the end of the nineteenth century. Two inside cylinders drive the leading coupled axle. (Reprinted from F. Pettigrew, 'Furness Railway Locomotives', *Proc. I.Mech.Eng.*, Pt. 3, 1901, Fig. 20, Plate 155.)

An American locomotive of about 1900 would be of the same general type and weight as the British engine but with outside cylinders. Both countries used inside- and outside-connected six- and eight-coupled small-wheeled engines for freight, and similar types were employed in Continental Europe and the Colonies. Everywhere the two-cylinder type was by far the most common, usually with the cranks set at right angles, so that reciprocation of pistons and associated masses imposed a to-and-fro force on the engine which had to be countered by adding, usually to the driving wheel, a balance weight in addition to that needed to counter the action of the revolving masses, such as crank pins. The rotation of this additional balance weight gave rise to a periodic pressure on the rail, known as 'hammer blow' or 'dynamic augment', which was responsible for much rail damage. In certain cases, as described in the next section, the action of this mass could cause the wheel to rise clear of the rail, such 'wheel bouncing' being extremely destructive of the permanent way. These undesirable features were always more prevalent with two-cylinder engines than with multi-cylinder types where some degree of self-balance could be achieved by using the reciprocation of one piston to counter another, but with a two-cylinder engine this could not be done without other problems arising. For example, if the cranks were set at 180 degrees, a two-cylinder locomotive might stop on 'dead centre' or sufficiently near it to be unable to start from rest. To avoid this, the cranks were usually at 90 degrees, which made addition of reciprocating balance necessary, and some hammer-blow inevitable.

Scientific Investigation of Rail Working by Locomotives

Engineers such as Robert Stephenson knew that a locomotive could be affected by out-of-balance forces, but knowledge was qualitative, rather than quantitative. Theories of balancing the reciprocating and revolving masses emerged in the 1840s, but it was not until the end of the nineteenth century that scientific studies began to find out how balancing technique influenced rail working. Insight into the theory and practice of locomotive balancing, and the effect on track about 1900 is offered by Prof. W.E. Dalby's paper 'The Balancing of Locomotives'[24] delivered before the Institute of Mechanical Engineers, November 1901, which clearly identifies the major phenomena which were to be intensively studied in the 1930s and 1940s. By 1900 theoretical argument, experience of railway operation and laboratory experiment had led engineers to recognize the link between correct balancing of locomotives and rail fracture, and had raised the question whether, in certain circumstances, the driving wheel might actually lift clear of the rail above a critical speed, thus violently impacting it when falling back. Some writers give the impression that 'wheel lift', as it came to be called, was not identified until very high-speed running, and investigation into rail fracture demonstrated its reality in the 1930s, but though the phenomenon was not thoroughly investigated and photographed until then, engineers in the nineteenth century were aware that it could happen, and probably did. Before discussing this, it is necessary to summarize what is meant by balancing, and Dalby's own definition is clear and brief:

> ... The moving masses of an engine may be divided into those which revolve with the crank shaft, and those which the crank shaft reciprocates ... principles applying to revolving mass may be made to apply to the reciprocating masses. It is only necessary to suppose that the reciprocating masses are transferred to their respective crank-pins, and to treat them there as a separate revolving system, the balance weights found being those which when reciprocated will balance the reciprocating masses ... in locomotive work it is almost the universal custom to balance the reciprocating masses by revolving masses placed in the wheels, the actual balance weight in the wheel being the resultant of the balance weights required for the revolving and reciprocating parts respectively. There is therefore no need to discriminate between the revolving and reciprocating parts in the process of finding balance weights. Having settled how much of the reciprocating parts it is desirable to balance, include it with the revolving masses at the crank-pin, and consider the whole as a revolving system. ...[25]

Deciding the percentage of the reciprocating weights to be balanced became a subject of great controversy, because the revolving balance weight chosen was largely responsible for destructive track working. The revolving masses such as the crank-pins, were always balanced, except in the most unusual circumstances. Weights added above that were termed overbalance, the art being to get the percentage sufficient to give steady running without destructive working of the rail; if the rotating masses were not balanced, the engine was said to be underbalanced.[26]

The use of revolving weights—'balance masses'—fitted to the engine's driving wheels is generally supposed to have originated on the Eastern Counties Railway, England, in 1845 when Fernihough employed them.[27] The fundamental theory is attributed to Nollau (1847) and Le Chatelier (1849), the latter conducting experiments on the Orleans Railway with an engine lifted clear of the rails and run up to 3 rps.[28] Later on models and wheel-balancing rigs were to come into use.[29] In 1855 Clark published a theory of balancing which raised the question as to how much of the reciprocating weight must be balanced.[30] At first, there was a tendency to balance all the reciprocating mass, but very great wear was experienced on the tyres where the 'hammer-blow' (see below) was transmitted.

> ... In some cases the tyres had been eaten into, so to speak, quite half an inch at this spot. These engines had had a certain amount of the balance-weights taken out, and not much further trouble had been experienced ...[31]

Such heavy wear, to which the early soft iron tyres were susceptible, led to 'flats', which in turn damaged the iron rail. Steel tyres and steel rail countered such tendencies but by no means eliminated them. Indeed, the problem of rail and tyre damage was to become acute in the twentieth century, with the increases in axle load, revolving and reciprocating weights, and rotational speed of the coupled wheels, this latter critical for inducing wheel-lift and impact loading.

In the 1850s there was a general tendency to overbalance too much, that is to counter an overlarge percentage of the reciprocating masses, which led to reports of excessive and unequal wear on tyres. But the cause was soon diagnosed and it became the rule to restrict overbalancing to about five-eighths of the reciprocating masses for an uncoupled engine, and about two-thirds or slightly more for a coupled engine. But the increase in size of reciprocating parts, and the rise in steam pressure, had brought 'hammer blow' to the order of 4 tons or more by 1900, and the whole question needed re-examination. By then, the early theories of Nollau, Le Chatelier and Clark had been extended by the laboratory investigations of workers like Osborne Reynolds, at Owens College, Manchester, who used suspended models of marine and locomotive engines to check disturbing forces.[32] That these forces—in particular, that pressure loosely termed 'hammer-blow'—could severely damage the rail was evident, though the exact details were then unclear. With hindsight, it seems certain that driving wheel lift had occurred:

> ... There was a case in Australia about which he was never satisfied. They tampered with the balancing of some of their engines, some of which were heavy, while some of their rails were light. Whether the

rails were of good steel or not he could not tell ... but (they) were regularly broken by certain engines. There was no doubt whatever that the rails were broken, and at distances where the hammer-blow would have an effect upon them, distances of from 15 to 20 ft (4·575–6·1 m) according to the diameter of the wheel of the engine. It certainly appeared to him to be very doubtful whether the hammer-blow could ever be so severe as to break a rail. ... The other engines properly balanced running over the same rails did not break them ...[33]

This engineer believed hammer-blow might bend a rail, but break only track which was defective. Others expressed puzzlement at what had happened, as laboratory tests indicated the rail should not fracture under maximum hammer-blow; evidently these engineers did not consider that the driving wheels were rising clear and falling back onto the rail. Yet this had been demonstrated in the laboratory at Purdue University, in the 1890s, where steam locomotives were tested whilst held stationary on rollers:

... Experiments were made with the locomotive at Purdue University (1894) by passing lengths of soft iron wire, about $\frac{1}{32}$ inch (0·793 mm) dia under the driving wheel, when running at considerable speed on friction wheels. The wire was flattened, but not to uniform thickness, and at high speeds it was of its full diameter for a short distance, showing that the wheel had actually lifted. ...[34]

Here then was a pointer to the cause of those lengths of rail fractured in places separated by multiples of the driving wheel circumference:

... With regard to the broken rail question ... there were absolute investigations made in America with ... small wheel coupled engines, by measuring the circumference of the wheel and by ... measuring where the engine broke the rail all along the road, in fact tracing the whole thing out ...[35]

The whole matter of balancing different types of locomotives, with two, three or four cylinders, as described by Dalby and others is too lengthy a matter to be included here; suffice to present the main results of the analysis and list the sources in the references. The phenomenon chiefly relevant to rail working, apart from frictional wear between tyre and rail, was the variation of rail pressure along the line of contact between wheel and rail.

In Article 10 of his paper, Dalby writes:

... The variation of the pressure between the wheel and the rail, caused by the vertical component of the centrifugal force due to the part of the balance weight concerned in balancing the reciprocating masses, is called the 'hammer blow'. This description of the effect does not describe what takes place very well, because the variation of the pressure is not sudden, but continuous, except in the extreme

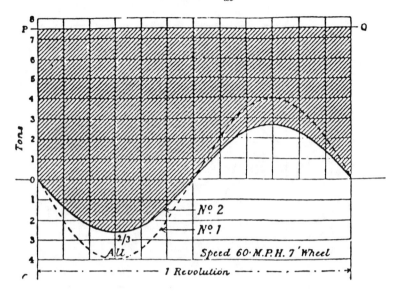

Figure 3. The figure shows the variation in rail pressure for one revolution of the driving wheels of a 'single driver' passenger engine. Curve No. 1 is for all the reciprocating masses balanced; Curve No. 2 is for two-thirds balanced. The static load per wheel is represented by PQ, $7\frac{1}{2}$ tons for this type of engine, so that the rail pressure at 60 mph (96 km/h), with an overbalancing of two-thirds, is given by the depth of the shaded figure.

case where the maximum value of the variation is greater than the weight on the wheel; in which case the wheel lifts for an instant, and coming down again gives the rail a true blow . . .[36]

In determining this pressure variation, the balance weight countering the reciprocating masses is crucial, for the variation of rail pressure per revolution changes with alteration of the degree of overbalance. A typical curve showing variation of rail pressure is illustrated in Fig. 3, drawn for a two-cylinder engine with one pair of 7 ft (2·135 m) diameter driving wheels.

This variation in rail pressure was intimately associated with wheel slip which, if occurring with wheel bounce, was particularly destructive of the rail. Even without wheel bounce, wheel slip was quite capable of hot-working the rail to such an extent that it was destroyed; while with wheel bounce the rail could be very rapidly cut through. Wheel bounce might occur without slipping, because it did not normally take place for both wheels on an axle at the same time, the driving cranks usually being out-of-phase by 90 degrees; thus the lifting wheel might be prevented from slipping if the others, to which it was connected through the axle and the coupling rods, provided the resistance to stop it. This is why 'single-

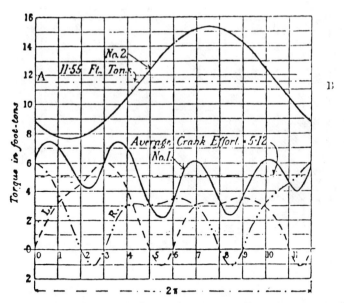

Figure 4. Driving torque, and couple resisting slipping, for a Lancashire & Yorkshire Railway 4-4-0 express passenger engine, at 65 mph (104 km/h), with two-thirds of the reciprocating masses balanced. (Reprinted from W.E. Dalby, 'The Balancing of Locomotives', *Proc. I.Mech.Eng.*, Nov. 1901.)

drivers', with only two driving wheels (one driving axle), were so much more prone to wheel slip than the four-coupled type with two driving axles coupled together.

Figure 4 shows the actual driving crank torque for a late nineteenth-century two-cylinder, inside-connected, 4-4-0 passenger engine. Also plotted is the couple resisting slipping.

Dalby remarks:

> ... The ordinates of curve No. 1 show the value of the driving ... torque; those of curve No. 2 the couple resisting slipping. It will be noticed how nearly the two values approach for crank position 1. If this had been a single engine, and with a little more steam, curve No. 1 would have cut No. 2, slipping being the inevitable result. In the case in question, the coupled wheels would come into play and prevent it ...

Yet if slip were avoided, the impact on the rail caused by wheel bounce or excessive hammer-blow (dynamic augment) could destroy the rail, even if it were free of those shatter cracks, caused by over-rapid cooling in the mill, which spread as transverse fissures under such impact.[37]

One way—which became standard—of reducing the variation in rail pressure in coupled engines was to distribute the balance weight intended to counter the reciprocating masses between the coupled wheels, instead of concentrating it on the main drivers to which the connecting rod big end was fixed. This enabled the magnitude of the variation of the rail pressure to be greatly reduced and Dalby quotes the case of a six-coupled goods engine in which distribution reduced it from 7·8 tons to 2·6 tons (77·65 kN to 25·9 kN). (Hammer-blow, in the British Imperial system, was expressed in 'tons', which was more accurately tons-force, 1 ton-force being 9·964 kN.) This was found to be the best way of protecting the track from hammer-blow and impact load due to wheel lift. The proportion of reciprocating mass balanced was about 2/3 about 1900, although this was not always distributed equally between the pairs of coupled wheels.

It must be realized that the balance masses, and heavy parts of the engine like the crank pins, were not in the same plane, even for one wheel, and that they were usually out of phase between the left-hand and the right-hand wheel of any pair. For very high-speed running, the disturbances resulting from this lack of balance across the engine had to be reduced by 'cross balancing', but this was more of a concern in the 1930s than in the nineteenth century.

Around 1900, locomotive engineers, while aware of the need for reducing the hammer-blow on rails or bridge members, could be confident of solving the associated problems without altering the basic form of Stephenson locomotive. They confronted no 'design impasse'. The basic engine was capable of very great development, with better balancing, as suggested by Dalby, being the best way of avoiding destructive track working. It is true that the more powerful types sometimes had three or four cylinders, the latter arrangement enabling considerable reduction of the reciprocating forces to be achieved, while more experimental engines introduced to minimize disturbing forces on engine and track met with little acceptance.[38]

Examples of locomotives specially designed to reduce loading on the rail include the eight-coupled freight engines of Ross Winans (Fig. 5), with the engine weight being spread over four axles and the cylinders horizontal. As the latter were above the leading coupled wheels, drive had to be through a jack-shaft and gears, to avoid inclining the cylinders—which would have been necessary had the piston rod been connected direct to one of the wheels. The engine running in the 1840s, like most then operating, was not provided with balance weights, not even for the revolving masses, and as a consequence tended to work track badly, though not as severely as if carried on the usual four-coupled wheels, with inclined cylinders directly connected.

One of the earliest schemes to produce a balanced locomotive, and one which was to exert considerable influence on succeeding attempts, was due to J.G. Bodmer who designed several balanced locomotives using four pistons moving in two cylinders. According to Ahrons,[39] J.G. Bodmer was one of the very first engineers to recognize the need to balance reciprocating masses and he took out a patent for an opposed-piston arrangement in 1834. Indeed, his name is used to identify a particular type of steam

Figure 5.   Vertical-boilered, eight-coupled engine designed by Ross Winans, and built by M.W. Baldwin for the Western Railroad of Massachusetts, 1844. The engine is sprung, the cylinders are horizontal, and the load is spread over four axles, but lack of reciprocating and rotating balance caused them to damage the track bed severely. Their performance was good, and some survived, with horizontal boilers fitted, to 1865. Weight was of the order of 23 tons. (Reprinted from E.P. Alexander, *Iron Horses*, Bonanza Books, New York, 1941, Fig. 17, p. 35.)

engine—the Bodmer engine—developed to provide high powers for stationary and marine application, and used in the 1830–50 period.[40] These opposed-piston machines of Bodmer were but one of several schemes involving oppositely-moving masses prepared in the 1830s and 1840s by engineers seeking to achieve steadier action of the locomotive. Bodmer's solution was particularly interesting in that he designed one piston rod to pass through the hollow rod of the other piston, moving in opposition within the same cylinder.

At least two of these engines, shown in Fig. 6, were built in 1845 for the South-Eastern and Brighton Railways; further details can be found in Ahrons and Walker.[41] As the plan view shows, the two cross-heads associated with each cylinder could not pass each other, so that one had to move to-and-fro ahead of the other. As the two pistons drove onto the same main crank axle, through a double-crank, one connecting rod, per pair, per cylinder, was shorter than the other. There is some evidence that the short rods tended to break. Both engines were inside-cylinder 2-2-2

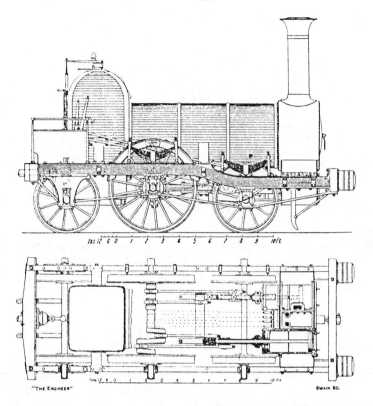

Figure 6. Bodmer's inside-cylinder, balanced engine, constructed for the South-Eastern and Brighton Railways in 1845. (Reprinted from E.L. Ahrons, *The British Steam Railway Locomotive 1825-1925*, Ian Allan, London, 1966, reprint of 1927 first edn, p. 61.)

types with a single four-crank axle. Each pair of pistons drove cranks at 180 degrees. According to Walker, Bodmer's engines aroused adverse criticisms, which has been put down to dislike of unusual practice and xenophobic dislike of Bodmer's Swiss nationality by his British colleagues.

Certainly eminent authorities, such as Major-General Pasley (Inspector-General of Railways), B. Cubitt, and J.J. Cudworth, testified that the engines were very steady running, and Ahrons writes that the Brighton engine ran as an opposed-piston machine until 1858 before conversion to a conventional two-cylinder, two-piston type. Walker provides details of an even more unusual design of Bodmer, for a 2-4-0 outside-cylinder type, prepared in 1844. Apparently several were built during 1845-6, some for the Manchester and Sheffield Railway by Sharp, Roberts & Co., of Manchester.

As shown in Fig. 7 they were remarkable in several ways. The two cylinders were external, and on each side the opposed pistons drove two connecting rods, two coupling rods and two external double-cranks. As

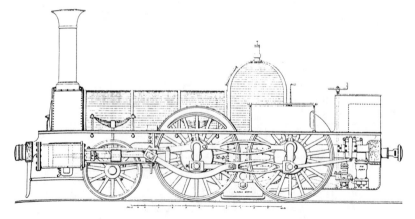

Figure 7. Bodmer's outside-cylinder, balanced engine designed in 1844, built 1845-6. The sledge brake is visible between the driving wheels, and the air pump for forcing air into the boiler with the feed water is beneath the footplate, to the rear of the trailing coupled wheel. (Reprinted from H.T. Walker, 'The Origin of the Balanced Locomotive', *Locomotive, Railway Carriage and Wagon Review (UK)*, 14 Feb. 1931, p. 43.)

the side elevation indicates, one of the connecting rods was bowed and joined the coupling rod some distance behind the leading double-crank. This enabled the connecting rod lengths and weights to be equalized, presumably to achieve practically perfect balance, though bowing a connecting rod seems a most undesirable practice.

As first built, these engines used sledge brakes, which were heavy shoes lowered onto the rail between the driving wheels, but these proved dangerous, being liable to foul points and cause derailments, and they were soon removed. The most remarkable feature was the combined water and air force pump driven from the rear crank, designed to force air, as well as water, into the boiler. According to Walker, Bodmer believed that the pressure of air would increase the 'Expansive action of the steam' with a consequent saving of fuel.

The coupling rods were double-bow-shaped to render them elastic and less likely to break as a result of shocks arising from defective permanent way. It is not clear how long these 2-4-0 engines ran, but the general principle of opposed pistons, working in the same cylinder, with co-axial piston rods, was not taken up generally as wear and leakage were probably excessive, resulting in high maintenance costs and power loss. Nonetheless, these engines did demonstrate that steady running could be achieved by using four pistons, geared to a crankshaft in such a way that the movement of reciprocating masses was self-balanced.

A more important design,[42] destined to prove very influential, was the 'Duplex', introduced by J. Haswell in 1862 for the Austrian State Railway (Fig. 8). Haswell employed Bodmer's basic notion, but simplified the

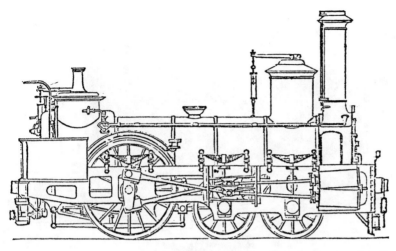

Figure 8. Haswell's four-cylinder 'Duplex' built in 1862 for the Austrian State
Railway.

design by assigning a cylinder to each of the two pistons on each side, thus
reducing wear and leakage, and enabling straight connecting rods of equal
size to be used. The 'Duplex' was a four-cylinder engine, all the cylinders
being outside the frames, arranged in pairs, one pair on each side. Each
pair consisted of one cylinder above another, driving the double-crank
outside the frame. There were thus two external double-cranks on the
engine, the two cranks of each pair being at 180 degrees, and the two
pairs being set at 90 degrees.

The reciprocating and rotating masses thus balanced on each side of
the engine, and there was no need for balance weights. It was the first
four-cylinder engine driving on one axle, and though this 4–2–0 express
locomotive ran very steadily, it was not reproduced, probably being
thought too complex. Instead, conventional designs were balanced with
weights. However, the 'Duplex' idea was continually revived in different
guises until the end of steam traction.

The balanced engine was an attempt to get rid of hammer-blow based
on Haswell's 'Duplex' principle, but usually employing four coupled
wheels with the cylinders in line across the engine and driving onto two
external double-cranks. A good example is that due to H.F. Shaw, built
in 1881 and shown in Fig. 9. It used the same principle as the 'Duplex'
but the two cylinders on each side were placed horizontally beside each
other, rather than one above the other, as with the Haswell engine. The
balanced-compound four-cylinder engine of Shaw was built by the Hinkley Locomotive Works in 1881 and weighed 37 tons without tender. On
each side of the engine, two cylinders, side by side, drove onto a double-
crank fixed to the leading coupled axle, so that two coupling rods, as well
as two connecting rods, were used each side, just as was done by Bodmer

Figure 9. The balanced-compound four-cylinder engine, built in 1881 by the Hinkley Locomotive Works to the designs of H.F. Shaw. (Reprinted from E.P. Alexander, *Iron Horses*, Bonanza Books, New York, 1941, Plate 75, p. 189.)

in his 1844 2-4-0. The arrangement enabled rotating and reciprocating masses to be sufficiently balanced to dispense with driving wheel weights and hence there was no hammer-blow. Though a success, it was not a system widely used, railways favouring the much simpler two-cylinder orthodox type, which avoided expensive double-cranks and extra rods.

The Shaw balanced compound-expansion engine was followed in the United States by 'Balanced-Compounds' of a more normal design, in which two of the cylinders were placed within the frames and two outside, all driving onto the leading axle, usually of a four-coupled engine. Though this type of layout became accepted in Europe, and was used with simple and compound expansion, it was abandoned in the United States around 1900, after which there arose an extreme prejudice against inside cranks, possibly through fear of crank failure or crank flexure; possibly as part of a quest for simplicity; and partly because the usual American bar-frame structure left less room for an inside cylinder, or pair of cylinders, than the British plate frame. The inside crank was not rejected by European designers and such differences between American and European practice need a deeper study by engineering historians than they have received so far.

There were periodic attempts to use four-cylinder compound-expansion systems to reduce rail stress down to the 1930s, and four simple-expansion cylinders to obtain some degree of self-balance of the reciprocating parts were also fairly widely used in Europe. In some cases the drive was divided between the first and second coupled axle, as on the Great Western Railway, where four-cylinder engines were introduced by G.J. Churchward in 1906. For the best results, however, all four cylinders had to drive one axle. This latter practice enabled more powerful engines to be used on lighter rail and trackwork than would otherwise have been the case, examples being the four-cylinder simple-expansion engines with drive to the leading axle which were extensively used in the Netherlands in the early twentieth century.[43]

Another design of the late nineteenth century which must be mentioned is the so-called 'Double-Single', consisting of two uncoupled driving axles, one (usually the rear) driven by the outside pair of cylinders, the other being driven by the inner cylinder or cylinders. The researches by O. Reynolds into the limitations to speed of stationary and locomotive engines not only developed a general theory of balancing for all kinds of reciprocating steam engine, but contained a study of coupling rod failure.[44] Reynolds clearly states that as locomotive speed was raised, the coupling rods were the part likely to fail first, and these studies—borne out by in-service mechanical failure in the 1870s and 1880s—caused some engineers to produce four-driving wheel locomotives without coupling rods. The success of 'Single-Driver' engines in high-speed service, and the advent of steam-sanding and air-sanding, in the late nineteenth century, led to several experiments with the 'Double-Single', sometimes using simple-expansion, sometimes with compound-expansion. The number of cylinders varied: four were used on the Drummond units of the London & South Western Railway and the American 'James Tolman'; three were employed on the Webb compound engines tried on the London & North Western Railway. Leaving off the coupling rods to save weight; to avoid having to balance them; to aid cornering; and to avoid failure at speed, resulted in a tendency for the 'Double-Singles' to slip excessively, wherever they operated, and though tried on the Pennsylvania RR, the London & North Western and the London & South Western, they proved unsuccessful, and the type was not continued in the twentieth century.[45] What is more, as the two sets of driving wheels could slip independently, the four cranks could get out of the phase in which the reciprocating masses provided self-balance.

It is all the more interesting, then, to note that when steam engine designers faced the problem of severe track working by very large engines in the 1930s and 1940s, they tried to reduce the resulting rail stresses by introducing a type (called the 'Duplex') based on the combined principles of the Haswell 'Duplex', the 'Double-Singles' and the Shaw balanced engine. There were even attempts to produce a quadruple-single, with four uncoupled driving axles powered by independent groups of multi-cylinder steam motors or small turbines.

In the 1930s there were widespread attempts to accelerate passenger and express freight services on the railways of the leading industrialized nations.[46] Some of the passenger services used diesel traction, but generally steam was employed, 4-4-2 and 4-6-2 locomotives being the most common.[47]

In the United States, many of the very fast services were operated by two-cylinder 4-6-2 engines, introduced in the early 1920s or even before the First World War, and greatly improved with new boilers, cylinders and draughting. As these engines were being run at far higher speeds than originally intended, particular care had to be given to balancing and cross balancing, especially as these accelerated services had resulted in severely damaged track, with 'hammer-blow' (dynamic augment), wheel bounce and slipping suggested as agents. The locomotive was not the sole cause of damage to track, and it was widely known that freight and passenger waggons could work the rail destructively,[48] but the growth of engine size, the increase in rail pressure forces, the occurrence of driving wheel bounce, and the greater incidence of 'rail burning' due to wheel slip, demanded some modification to engine design if the permanent way was to be protected from an excessive degree of injury. Both laboratory and trackside investigation was conducted, furnishing an exemplar of that fruitful co-operation between academic institutions and industries which has become increasingly necessary for commercial progress in the twentieth century.

The Urbana Investigations of the Impact Loading of Rail

Of particular importance were the series of tests on the impact stressing of a freely-supported beam carried out in the Materials Testing Laboratory, University of Illinois, Urbana, by R.N. Arnold. As described in an earlier paper,[49] this department rendered great services to industry by fostering a range of improvements in rail technology, from manufacturing techniques to in-service maintenance methods. The aim of R.N. Arnold's work was:

> ... (to review) ... various theories used in estimating impact effects on a freely supported beam, and ... (errors) ... which may be introduced if the assumption is made that elastic forms are similar under dynamic and static loading ...[50]

The freely-supported beam was an 8 ft (2·44 m), 112 lb/yd (55·58 kg/m) length of standard Stevens–Vignoles railway rail placed under a drop hammer (Fig. 10). An electric motor raised the weights to the desired height, up to 5 ft (1·524 m), the combined masses of load and striker being either 50, 98, 239 and 470 lb (27·68, 44·5, 108·5 and 213·4 kg).

These weights were held at the required height by the electromagnet, D, which let them fall when the current was switched off. The striker, C, was machined to the standard radius of the American car wheel, $16\frac{1}{2}$ inches (419·3 mm).

Each specimen length of rail was supported resting on the roller, A, with the rail end held down by four set screws to approximate to the condition of free support. Experimental aims were to investigate the variation in bending stresses during impact in the centre of a freely-supported

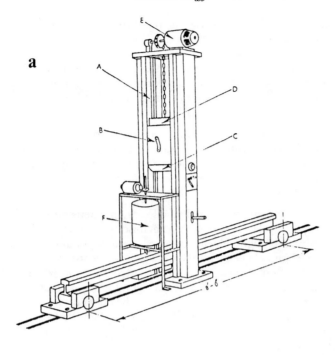

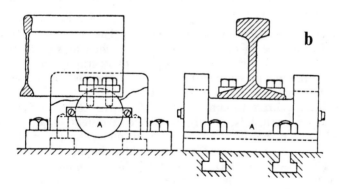

Figure 10. Drop hammer used in the Urbana impact tests. (a) Experimental apparatus used by R.N. Arnold at Urbana to investigate the impact loading of Stevens-Vignoles rail. The weight, B, is lifted in the guides, A, by the electric motor, E. Switching off the electromagnet, D, releases the weight, B, and the striker, C, which impact the rail. The rotating drum, F, records the space-time diagrams for the heaviest falling weight. (b) Details of end support for the rail. (Reprinted from R.N. Arnold, 'Impact Stresses in a Freely Supported Beam', *Proc. I.Mech.E.*, Vol. 137, 1937.)

railway rail; to study load variation on a freely-supported railway rail under central impacts; and to identify the practical significance of the experimental results.

Impact stresses were the prime concern and were measured in two ways, one crude and the other more accurate. The crude method was intended to indicate directly the maximum load between the striker and the rail during impact:

> ... This was done by inserting a very thin piece of paper between the two bodies during impact and measuring the area of the ellipse of contact obtained on the paper. By repeating the procedure with the same striker under known static loading an estimate of the maximum impact loads was obtained ...[51]

But the main investigation used small scratch extensometers, similar to those developed by De Forest at Massachusetts Institute of Technology, modified to record high-velocity impacts. Arnold comments that at the time of the Urbana experiments, there were few records of dynamic loading stresses, but much theoretical work had been done which could be related to laboratory models of in-service conditions, such as the analysis by Timoshenko[52] of a sphere impacting a beam and the deflections of railway track under dynamic load. He refers to Timoshenko and Langer (1932), who treated the case of a wheel rolling through a low spot, or depression in the rail, but points out that:

> ... They derived a dynamic (loading) factor from consideration of deflexion, but assumed that the wheel followed the curve. This is not necessarily true for all speeds ...[53]

A valuable technique was demonstrated by Tuzi and Nishida[54] in 1936 who subjected a beam to central impact and then took photographs of the photo-elastic fringes using high-speed photography—a study seemingly indicating the presence of the sub-impacts detected by Arnold.

Undoubtedly the study of locomotive hammer-blow effects on bridges by the Bridge Stress Committee,[55] 1928, of the Dept. of Scientific & Industrial Research, UK, greatly stimulated investigation, as well as providing deflection records of a series of spans under the action of different types of engine. The strains due to sudden impact on girders were recorded, the results supporting Mason's 1938 developments of the theory of transverse impact of a beam,[56] based on earlier work by Timoshenko,[57] Cox[58] and St Venant.[59] Arnold treated the rail specimen as a flexible beam, subject to in-service loading by locomotive and car wheels, due to a number of causes such as running into a low spot; a tyre with a 'flat'; closed springs; the sudden surging of liquid in tank cars; hammer-blow, and—most severe—driving wheel lift, coupled with slipping. Arnold's scratch extensometer records are very similar in form to those taken from a stretch of main line, as shown in Fig. 11, thus justifying application of lessons learned in the laboratory to in-service problems.

As Arnold remarks:

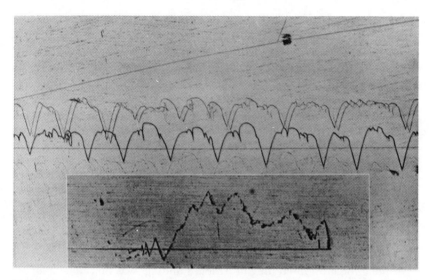

Figure 11. Bending strain trace due to a train running over a 'low spot' in a rail, recorded on a main line railway, USA. The inset shows the trace for one of R.N. Arnold's laboratory tests, in which the specimen rail is struck by the 213·4 kg mass. (Reprinted from R.N. Arnold, 'Impact Stresses in a Freely Supported Beam', *Proc. I.Mech.Eng.* (*UK*), Vol. 137, 1937, pp. 217–81, Plate 3.)

... (The figure) shows a scratch record of the bending strain at the joint of a rail, during the passing of a train. It was taken on a main line railroad in the United States and in form is not very different from the records of bending strain obtained in the present investigations. ... Theoretically a railway track is comparable to a beam on an elastic foundation, but this does not seriously affect comparison with a freely supported beam. Both seem to show similar strain variation, which in the case of the falling weight is directly due to the influence of sub-impacts. It therefore seems evident that the impacts of wheels on railway tracks are very similar to those studied experimentally ...[60]

Arnold drilled small holes along the centre of the lower rail flange, and used them to fix extensometers for recording strain in each 2 inch (50·8 mm) length of one half of the test specimen.

The first tests showed clearly that:

... the strain variation at the centre of the rail ... (was) of a complex nature, quite unlike that which would be obtained from static loading ... it is evident that a portion of the rail at each end bends initially convex upwards during the impact. The elastic curve taken up by the rail at that instant is therefore not equivalent to the static curve so often assumed for calculating impact stresses ...[61]

A major feature of these Urbana investigations was the identification of sub-impacts, which combined to form the main impact, the number of component collisions increasing with the weight of the striker.

> ... The sequence of events during the impact ... may be described in the following manner: Sub-impact 1: (a) The weight strikes the rail; a large reaction is set up, and the velocity of the weight decreases. (b) The centre of the rail, being subjected to a high load, accelerates quickly and attains a velocity greater than the decreased velocity of the weight. Contact ceases. (c) The rail deflects, converting its kinetic energy to strain energy, comes to rest, and reverses its velocity. Meanwhile the weight is approaching the rail at its reduced velocity. Sub-impact 2: The weight again strikes the rail, having its velocity further reduced and again accelerating the rail away from it. This procedure recurs until the velocity of the weight is reversed and becomes such that the rail cannot again overtake the weight. These sub-impacts all occur during a fraction of a second ...[62]

Among the most interesting of Arnold's findings was that the maximum impact load (web strain) for the small weight used in the tests was not very much smaller than for the greatest (50 lb (27·68 kg) vs 470 lb (213·4 kg)), and that for equal impact velocity, weights above 300 lb (136·2 kg) did not much affect the maximum reactive load set up in the deflecting rail. This fits in with trackside tests, during which it was discovered that engines with a low axle load could be every bit as destructive of rail as those whose axle load was much higher.[63]

Even at low speed, light locomotives could be more destructive of rails than heavy high-speed engines, provided wheel lift occurred, or hammer-blow was high; and if the driving wheels were small, this lift could occur at a moderate speed.

This still implied that higher operating speeds brought increased problems, as the greater velocities made wheel lift more likely especially for older engines designed for lower speeds.

As Arnold comments:

> ... In the case of a beam, it appears that the kinetic energy of a large weight is dissipated during a large number of sub-impacts. The maximum load exerted, however, may not be far in excess of that exerted by a much smaller weight which dissipates all its energy in one sub-impact ...[64]

In the concluding part of his paper, Arnold discusses the practical application of his findings, being of little doubt that they were of use in solving problems arising from railway operation. For example, his work, like earlier investigations at Urbana,[65] showed that impact stresses due to wheel load could be high, even when the bending moments in the rail were low.

If the rail contained transverse fissures, such as originated in over-rapid cooling at the rolling-mill, fracture could result. Of course the in-service conditions were considerably more complex than those investigated in the laboratory, for—as Prof. H.F. Moore and Prof. H.R. Thomas of Urbana pointed out[66]—in actual service the condition at any point in the rail was decided not only by the action of the nearest wheel, but by adjacent wheels as well. Furthermore, the rail supports, track structure and track bed had considerable influence, as the engineers of the 1830s had come to appreciate.[67] Climate was also important, for track flexibility changed greatly if, for example, the moisture in the ballast froze. Moore and Thomas thought that this flexibility would reduce in-service loads below those predicted by Arnold, though in reply the latter stressed that because actual track was not more flexible than the experimental beam, as his critics supposed, the impact loads in service could equal those predicted on the basis of laboratory experiment, namely 2·5 times the magnitudes calculated by 'static equivalence' methods. Moore and Thomas thought that:

> ... When the flexibility of the support of a railroad rail was considered it seemed unlikely that the wheel loads and resulting stresses would be as high as was indicated by the factor 2·5 ... A railway rail was supported on yielding ties, ballast, and road bed ... the yielding support would lengthen the time required for the rail to acquire the instantaneous vertical velocity of the 'falling' wheel, that is the deceleration of the wheel would be reduced below the value found in tests made with apparatus as rigid as that used by the author. ... Just to make the picture still more complicated, it was quite probable that the stiffness of track support was greater under rapidly applied load than under slowly applied load ...[68]

Certainly the phenomenon was a complex one, and in the late 1930s there was a scarcity of information about what did occur, say, when a driving wheel lifted. How did the compressed springs accelerate the unsprung parts downwards, and what was the action of steam pressure, transmitted through the connecting rod, on this accelerating mass? Though thorough investigations were begun, these questions were never answered completely before steam traction was recognized as obsolete.

Investigations of Wheel Lift and Engine Balancing

Following the Urbana laboratory investigations, and the positive identification of in-service driving wheel bounce in 1937,[69] extensive trackside tests were undertaken in both the United Kingdom and the United States to study the impact loading of rail, and the best ways of balancing locomotives to prevent track damage. The situation in Britain around 1941 is stated in the paper *Balancing of Locomotive Reciprocating Parts* by E.S. Cox, who sums up the problem from the locomotive engineer's point of view:

> ... it is universally agreed that revolving masses should be completely balanced; but there has been, and still remains, considerable

> divergence of opinion as to how the reciprocating masses are to be dealt with ...[70]

Left unbalanced, particularly on a two-cylinder engine, these caused a to and fro periodic force plus a nosing couple; but when balanced with driving wheel masses, they led to that variation in pressure between tread and rail which for decades had been called 'hammer-blow'.

> ... Two conflicting needs have to be satisfied: that of the mechanical engineer who wishes to avoid all longitudinal and transverse disturbing forces on the locomotive itself, and that of the civil engineer who wishes to avoid all hammer blow effects on rails and bridges ...[71]

Cox mentions that early locomotive engineers, like the Stephensons, were compelled to balance the reciprocating parts which, in action, were damaging the very light engines of that time. Opposed pistons were tried at first, but after 1845, masses placed in the driving and coupled wheels became the standard, twentieth-century practice being to balance about two-thirds of the reciprocating weights.

The British Bridge Stress Committee report of 1928 recommended that locomotives should be designed so that at 5 rps, the driving axle hammer-blow was not greater than one-quarter the static load, with 5 tons (49·82 kN) maximum for one axle, and 12½ tons (124·5 kN) maximum for the whole engine, a recommendation followed in new construction during the 1930s.

(Hammer-blow, in the British Imperial System of Units as then employed, was usually expressed in tons-force, with 1 ton-force being 9·964 kN.)

These recommendations were to be reviewed, at the end of the 1930s, when the speeding up of services had increased express locomotive wheel speed from 5 to 8 rps, necessitating a general re-examination of balancing practice. This had varied considerably from railway to railway and from engine type to engine type, being in part dependent on how the balance weights were distributed and the number of cylinders.[72] Certainly multi-cylinder engines were easier to balance and therefore less heavy on track, Cox quoting the example of a large four-cylinder Class 7 4–6–2 engine exerting a hammer-blow at 8 rps no more than that exerted at 5 rps by the much smaller three-cylinder Class 5X 4–6–0.[73] Cox summarized the results of an investigation into speed and wheel lift, carried out at a time when very high speeds—up to 125 mph (201 km/h)—were being attained on test in Europe and the United States, and when 100 mph (160 km/h) was timetabled for the fastest trains. These speeds were not always reached by engines with large coupled wheels, and there was an increase of the times when the centrifugal force associated with the balance mass added to counter reciprocating weights exceeded the driving wheel static load, leading to wheel lift. A large wheel diameter in 1900 was about 80 in (2032 mm), but by 1930, 70 in (1778 mm) or even less was common on express engines. Generally, as speeds rose after 1900, the coupled wheel diameter decreased from above 84 in (2134 mm), with 90 in (2286 mm) common, to approximately 74 in (1880 mm) for a fast engine by 1940. As Cox comments:

> ... Modern free-running valve gears allow even mixed traffic engines with 6-foot wheels to reach and exceed 90 m.p.h. in ordinary service. A rotating speed of 8 r.p.s., which represents 103 m.p.h. with 6 foot wheels, and 115 m.p.h. with a 6 ft. 9 in. wheel, is therefore a definite possibility and will have to be allowed for in future ...[74]

American practice was somewhat different at this time. Less attention was given to cross balancing, and two-cylinder engines were almost universal apart from articulated units.

Reciprocating masses were much heavier than in Europe and general practice was to balance only low fractions of the reciprocating weights. Speeds were as high as in Britain and, therefore, hammer-blow and wheel lift were equally great problems in the United States as in Europe, despite the low degree of overbalance:

> ... In spite of occasional outbreaks of broken and bent rails, the effect of this phenomenon passed for many years comparatively unnoticed, the great strength of the track and bridges absorbing the vertical forces. With higher speeds, however, and particularly where slipping of the wheels occurred, the wheels have been found actually to leave the track by an appreciable amount, and to bounce up and down, delivering heavy blows to the rails. ... In this country no authentic observation had been made of similar occurrences, but from time to time rather mysterious cases of bent rails occurred— evidence which seemed to connect with the slipping of the coupled wheels of locomotives and an investigation was put in hand ...[75]

This investigation, by the London, Midland & Scottish Railway Research Dept., used three two-cylinder 4-6-0 engines[76] having their reciprocating masses balanced at 66·6, 50 and 30 per cent.

The bull-head track was greased to induce slipping, with a high-speed camera photographing wheel lift, rotational speed and the movement of the engine forward along the rail. Slipping speeds equivalent to 110 mph (177 km/h) were achieved, though the forward speed of the engines was between 10 and 18 mph (16·1 to 28·9 km/h). The results, as presented by Cox, are reproduced in Table 1.

The wheel bounce damaged the track severely, the rail being bent with the ends lifted above the middle portion—the same form as was discovered by Arnold at Urbana.[77] Severe damage extended to the track bed, and rail, sleepers and ballast needed reconstructing afterwards (Fig. 12). It was reported that:

> ... In all the tests the amplitude of the vertical oscillation of the driving wheels was greater than that of the leading or trailing wheels, although the hammer blows are the same. This appears to be due to

TABLE 1 Results of wheel-lifting tests

	5043		5464		5406	
Engine No.	5043		5464		5406	
Percentage of reciprocating parts balanced	66·6		50		30	
Slipping speed, mph	103		104		99	
Maximum lift of driving wheel (inches)	2·4		0·4		No appreciable lift	
Engine oscillations	Nothing abnormal		Moderate oscillations		Excessive oscillations	
Hammer-blow:	5 rps	103 mph = 8 rps	5 rps	104 mph = 8 rps	5 rps	99 mph = 7·7 rps
Per wheel (leading, driving, and trailing) (tons)	3·84	9·82	2·95	7·55	1·76	4·18
Per axle (leading, driving, and trailing) (tons)	4·28	11·00	3·50	8·95	2·10	4·98
Total engine blow per rail (tons)	10·36	26·60	7·59	19·40	4·04	9·60
Whole engine blow (tons)	11·52	29·55	9·03	23·10	4·83	11·48

This table shows the results of the wheel-lifting tests carried out by the London, Midland & Scottish Railway, using three two-cylinder, outside-connected 4–6–0 locomotives, with different degrees of overbalance. The third column results show that two-cylinder engines must carry more than 30 per cent overbalance if excessive to-and-fro oscillations are to be avoided.

Reprinted from E.S. Cox, 'Balancing of Locomotive Reciprocating Parts', *Proc. I.Mech.E.*, Vol. 146, 1941.

Figure 12. Bent rail taken out from chairs after wheel slip tests on London Midland & Scottish Railway. The whitewashed edge marks the upper side of the rail head. (Reprinted from E.S. Cox, 'Balancing of Locomotive Reciprocating Parts', *Proc. I.Mech.E.*, Vol. 146, 1941.)

the position of the driving wheels near the centre of the deflected length of track ...[78]

Figure 13 shows a frame from the film made during the tests with the driving wheel (to which the main connecting rod was fixed) lifting off the rail through an amplitude greater than the coupled wheels. Overall conclusions were that wheel bounce was a forced vibration caused by the out-of-balance forces; that the coupled wheels lifted at lower speeds than theory suggested; that bouncing and damage diminished with hammer-blow magnitude, no lift occurring if less than 30 per cent of the reciprocating masses were balanced; that at least 30 per cent of the reciprocating masses had to be balanced, for two-cylinder engines, if undue longitudinal oscillations and nosing were to be avoided; and that the condition of track had little effect on the incidence of wheel lift. As a consequence, it was suggested that engines similar to those tested—a very common type—should have 50 per cent of the reciprocating mass balanced, rather than the usual two-thirds. Figure 14 shows the vertical motion of wheel and rail plotted against hammer-blow phase. Cox went on to investigate other problems, such as fore-and-aft oscillation and nosing, referring to ten major types of disturbance set up in a moving engine. These were extensively reported elsewhere.

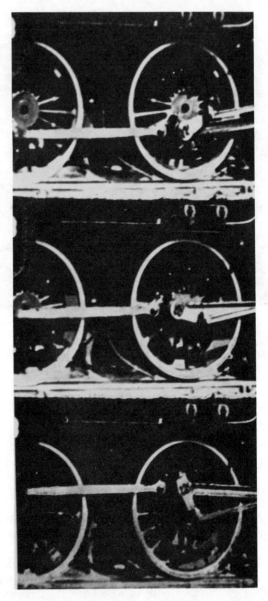

Figure 13. Frames from film of wheel slip investigation, London Midland & Scottish Railway. Driving and trailing coupled wheel of 4-6-0 engine shown. Direction of travel, left to right. The driving wheel has clearly lifted off the rail. (Reprinted from E.S. Cox, 'Balancing of Locomotive Reciprocating Parts', *Proc. I.Mech.E.*, Vol. 146, 1941.)

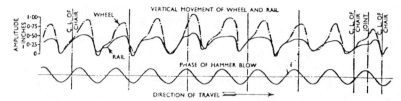

Engine No. 5043 : 66 per cent reciprocating parts balanced. Mean slipping speed, 102 m.p.h.

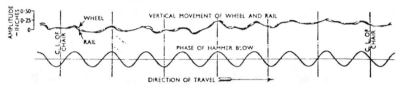

Engine No. 5406 : 30 per cent reciprocating parts balanced. Mean slipping speed, 97 m.p.h.

Figure 14. Diagram to illustrate driving wheel bounce. Vertical movement of the main driving wheel, and rail displacement, plotted against hammer-blow phase, as measured in wheel lift tests, London, Midland & Scottish Railway, United Kingdom, 1941. Both engines were two-cylinder, outside-connected 4-6-0 types, with a main driving wheel axle load of 18·2 tons, and driving wheels of 72 in dia. (1,829 mm). The upper trace shows driving wheel lift of approx. magnitude 13 mm, at 8 rps when the wheel hammer-blow is 9·82 tons, exceeding the wheel static load of 9·1 tons. See Table 1. (Reprinted from E.S. Cox, 'Balancing of Locomotive Reciprocating Parts', *Proc. I.Mech.E.*, Vol. 146, 1941.)

Cox drew nine major conclusions, chief among which were that wheel bounce could occur under British operating conditions, and that the hammer-blow on rail, for the recommended percentage of reciprocating masses balanced, was much higher than the 1928 Bridge Stress Committee visualized. It was suggested that three- and four-cylinder engines could be balanced with respect to longitudinal forces without adding balance weights to the driving wheels, and that certain classes[79] were already operating with zero reciprocating balance. For these engines there was zero hammer-blow; but for two-cylinder types, by far the most common in any country, and universal in the United States apart from articulated freight engines, some degree of reciprocating balance was needed, and hence some hammer-blow would be exerted on the rail. Cox recommended that not less than 40 per cent be balanced for the larger British two-cylinder type, weighing 65–75 tons without tender, and in the final British steam designs, the Standard Classes, introduced after 1951 when, in fact, steam was nearly obsolete, the figure chosen was 40 per cent for some types, 50 per cent for others. Tables of hammer-blow are included in Cox's description of these engines.[80]

There was never any universally accepted opinion about the best method of balancing engines. Colam and Watson,[81] in a paper published with Cox's, argue that it might be abolished altogether, and that 'overbalance' is in every way objectionable, despite its tendency to counter nosing, which causes the engine to strike first one rail and then the other. Colam and Watson referred to actual operating conditions for examples of engines performing apparently satisfactorily in a state of 'unorthodox' balance, such conditions being revealed by the British Bridge Committee's investigations of 1928, and the Indian Bridge Committee's survey of 1926.[82] They quote overbalance* ranging from +51 per cent to −15 per cent, the latter figure indicating: '... that even the revolving weights were not completely balanced ... These engines had given eminently satisfactory service for many years ... no-one ... had any complaints about them ...'[83]

Other investigations in India identified overbalance ranging from +80 per cent to −50 per cent:

> ... In the latter case, which was found on a South Indian Railway 4-6-0 locomotive, in no wheel were the rotating masses balanced. Far from the reciprocating masses being partly balanced, their weight of 403 lb. was virtually increased to 600 lb.; but again there were no evil results ...[84]

As a result of their Indian experiences, the authors argued that the percentage of the reciprocating weights balanced could either be much less than that generally recommended or overbalance could be omitted altogether. They conducted extensive trials to investigate the effect of different types of engine on rail and bridge member, concluding that:

> ... No trouble has been experienced which can in any way be connected with the lack of overbalance ... and no difference can be detected between these engines and engines normally balanced ... The authors consider that a prima facie case for abolishing overbalance has been established, and that if locomotive engineers wish to continue this practice, they must prove that it is necessary. It is certainly expensive. ...[85]

These recommendations were not generally followed. Overbalance was only omitted on multi-cylinder engines where the cranks had been set so that the reciprocating masses balanced each other; elsewhere, especially on two-cylinder engines, some was deemed necessary. It should also be realized, as Colam and Watson admit, that underbalanced engines would impact the rail as the rotating masses would not be balanced; thus there would be a heavy blow on the rail due to the action of crank pins, connecting rod big-ends and coupling rods.

That they suggest no undue harm resulted from the extreme overbalance and underbalance they discovered might be due to the low operating speeds in India, especially on the metre gauge system. Certainly their findings were not applicable to the very high-speed operations becoming common in Europe and the United States.

Some engine types, including the British War Department 'Austerity'

* If expressed in lb-weight, or lb-force, 1 lb-force = 4·448 N.

class 2-8-0 two-cylinder machines dating from the 1939–45 war, were built without overbalance, being constructed for low-speed service in wartime conditions.[86]

Rail-Stress Measurements and Locomotive Counterbalancing

In the 1930s there was a general speeding up of rail services in Europe and the United States which made closer study of rail damage by locomotive action necessary. In the United States, higher speeds were obtained on some systems by using lightweight diesel units, but generally this mode of traction was insufficiently advanced, before 1940, to warrant a rapid switch to it, or to provide reliable diesel-electric locomotives for working the heaviest, fastest trains. Though diesel-electric traction had been shown to be a potentially useful system as early as 1934, by the demonstration units of the Electro-Motive Division of General Motors, improvements to the operation of steam engines over long distances enabled the latter to survive into the early 1950s, and during these 20 years or so, there was a series of fruitful tests on full-sized track and rolling stock, aimed at relating track working to engine design, and developing the theory relevant to the phenomenon.

There was such variation in balancing practice, and a lack of a universally agreed theory, that tests were essential for clarifying the matter. In this respect, a particularly useful series of tests were those conducted by the American Association of Railroads, for the Chicago & North Western Railway, the results being published in the United States and the United Kingdom between 1941[87] and 1948.[88] The original experiments were the consequence of that railway company introducing a high-speed service from Chicago to St Paul and Minneapolis. Because it was timetabled to cover the 409 miles from Chicago to St Paul in 390 minutes, it was called the '400'.

Like similar expresses on other railways, such as the Union Pacific 'Forty-Niner' between Chicago and San Francisco, and the Pennsylvania 'Jeffersonian' between New York and St Louis, the '400' was originally operated by uprated 4-6-2 steam engines, in this case class E3 two-cylinder types, greatly improved since first construction by balancing, cross balancing, better steam distribution and more effective draughting:

> ... The question of the degree of improvement achieved by such changes—the reduction of the stresses induced in rails at various running speeds, and the extent to which safe operating speeds may be increased—is one that can be answered with less assurance by mathematical computation than by direct experiment. The C. & N. W. Rly ... (has) ... provided facilities for a difficult but most instructive series of rail-stress measurements. ...[89]

Although many of these trains, the '400' included, were diesel or electric powered by 1941, sufficient steam power remained in service to warrant developing these initial tests into a whole series, the results of which had considerable influence on the final phase of steam engine design in the United States. These tests were to show that hammer-blow, 'dynamic augment' to use the US term, arising from overbalance, could be related to the stress in the foot of the rail and that:

> ... the measured rail-stresses under all the coupled wheels, except the main driver, agreed closely with the stresses calculated from the nominal wheel loads and counterbalance data. Under the main drivers, the agreement between measured and calculated stress was good except at the higher speeds, the disparity being attributed to the secondary effects of the piston thrust on the main connecting rod, and inertia of the reciprocating masses ...[90]

The tests clarified balancing practice, and led to further trials with a greater range of engine types, investigating how other components—such as the bogies—worked the rail.

Very useful results were obtained. For example, rebalancing certain 2–8–2 freight engines reduced rail stress in 100 lb/yd (49·61 kg/m) track to such a degree that instead of being limited to 50 mph (80·45 km/h), speed was then '... restricted only by the capabilities of the engine ...' This better balancing reduced the likelihood of wheel bounce and slipping, which earlier tests had revealed as being an even more complex phenomenon than previously thought:

> ... H.R. Clarke and K. Cartwright assert that locomotive driving wheels frequently attain revolving speeds equivalent to 25 to 30 m.p.h. higher than the actual running speed, and that, once such slipping starts, the engine may travel as much as a mile before the wheels resume normal behaviour.[91]

It should be noted that the tests in England, reported by Cox, involved slipping speeds very much higher than the forward speed, with rotational speeds corresponding to over 100 mph (160 km/h) occurring at a forward speed of 20 mph (32 km/h).

These latter tests were for slow forward speed slip, and when high forward speed slip took place, say at 90 mph (144 km/h), the difference between speed corresponding to revolutions and forward motion was much less.

Other design improvements encouraged by the 1941 tests included fitting engines with driving and coupled wheel centres less liable to distortion than the traditional spoked pattern, and which were lighter, a common pattern being the double-disk construction, linked by tubular stays. This type had a considerable effect on improving engine riding, for careful measurements had shown[92] that driving wheels departed from the round by a quantity described as 'by no means negligible', the phenomenon being due to uneven tyre wear caused by hammer-blow, and distortion of the orthodox spoked centre.

Some of the tests were conducted using this type of wheel, with new tyres to ensure negligible departure from the round, fitted to an E3 4–6–2 express passenger engine of the type used on the high-speed '400' service.

In the last generation of American steam engines, it was virtually a standard feature.

The same stretch of track was used for the 1941 series of tests and for the two later series of 1941/2. The tests involved fifteen types of engine from different railways and seven classes of wheel arrangement. Basically the methods of investigation were the same, as far as rail-stress measurement was concerned, though the later trials analysed the different modes of vibration of different parts of the engines, at various speeds, more thoroughly than in earlier experiments. The general method was to measure the longitudinal stress along the centre line of the rail base and to use this to calculate the vertical load applied to the rail through the wheel, and hence the hammer-blow arising from overbalance. Thus a preliminary step, as the report emphasized, was to survey the load–stress relationship for a rail, which meant finding the modulus of the rail support defined as the distributed force per unit length of rail needed to produce unit vertical depression. In the American tests this was assessed by applying light and heavy static loads to the track using the end wheels of a passenger coach (13,050 lb/wheel, 58,480 N/wheel), and the rear coupled wheels of a 4–8–2 engine (32,235 lb/wheel, 143,800 N/wheel):

> ... The wheels were successively spotted over each strain gauge, and the rail depression at the loading point and at three sleeper spaces beyond were measured by precise levelling to the nearest hundredth of an inch. ...[93]

The average depression differed from one rail to the other, 0·137 in (3·48 mm) to 0·117 in (2·97 mm) giving a mean of 0·127 in (3·225 mm) for the whole track corresponding, for the first series of tests, to an average modulus of 3,000 lb per linear inch per inch of rail depression (20·7 N per linear mm per mm of rail depression).

When the modulus was reassessed in the later series of tests it was found to be 1,900 lb per linear inch per inch of rail depression (13·1 N per linear mm per mm), the difference being due to temperature: the first tests were conducted in January when the track bed was frozen and the second series in summer and autumn when the bed was more flexible.

For the more comprehensive later series, made outside the winter period, the figure of 1,900 lb per linear inch per inch of rail depression was used (13·1 N per mm per mm).

In the first series of tests the rail used was of the section shown in Fig. 15, being classified as RA-A 10020 section, Stevens–Vignoles, 100 lb/yd (49·6 kg/m), laid in standard 39 ft (11·89 m) lengths, with staggered joints joined by two-bolt fishplates. It rested on tie plates, 7 in × 10½ in (177·8 × 266·6 mm), mounted on hardwood sleepers, 24 to the rail length, the metals being canted inward at 1 in 40. For the early tests the rail used was described as being in excellent condition, rolled in 1936 by the Illinois Steel Company, and comparatively little worn, the wear indicated by the

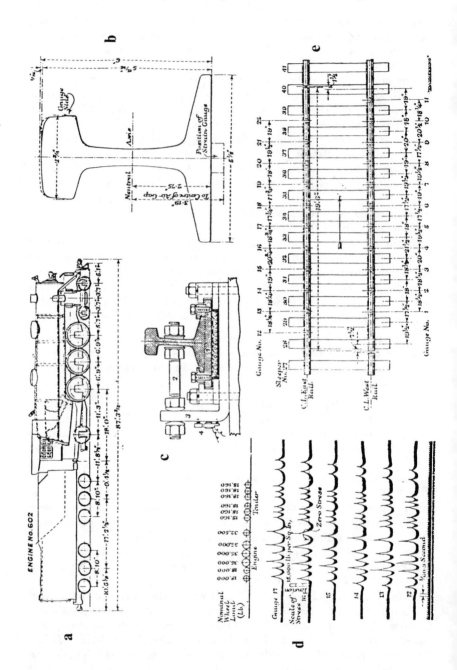

dotted profile in Fig. 15b. The strain gauges were fitted as shown in Figs 15c and 15e, the arrow in the latter giving the direction of travel of test trains.

The 'hammer-blows' were measured using magnetic strain gauges, each made up of two components fixed to the rail 2·25 in (57·15 mm) apart. One component was an electromagnet with a 'U' shaped core; the other the armature. Variation in the air gap between the two changed the system's reluctance, enabling length changes to be interpreted as resistance changes to the 2,000 cps A.C. passed through the electromagnet windings. The strain-gauge winding was part of a Wheatstone bridge circuit, the 'out-of-balance' current being fed to an oscillograph and photographic recorder. In obtaining the rail stresses due to hammer-blow from these readings, it was necessary to allow for stresses arising from temperature changes during test.

The report summarizes the general procedure as follows:

> ... to take (with each of the test locomotives) ... three sets of records with the locomotive passing at about 5 mph over the gauged length of track; and secondly to repeat these observations for the locomotive drawing a passenger train of suitable makeup at two or three higher speeds, runs being made at the maximum speed desired for operation in normal service, and at the highest speed that could be attained over the test length of track. In order to gain information on the effect of the steam load on the pistons upon the forces delivered to the track, half of the high-speed runs were made with the engine working normally, and the other half with the engine 'drifting' (steam shut off). From the 5 mph records, the average of the stresses measured by all the eleven gauges affected was determined for each wheel of the locomotive, and the mean results for the two rails compared with the theoretical stress due to the nominal static load imposed by the wheel. At crawling speed, the dynamic augment of load due to overbalance on the coupled wheels is negligible ...[94]

Five engines took part in the tests, four of the 2-8-2 freight type and one class E3 4-6-2, No. 602. The latter, Fig. 15a, was the heaviest of the five, having 72,000 lb (320·1 kN) on the leading truck; 210,000 lb total (934 kN) on the three driving axles; 65,000 lb (289 kN) on the single axle

Figure 15. Counterbalancing effects and rail-stress measurements, Chicago & North Western Railway, USA, 1940-1. (a) Dimensions of CNWR 4-6-2, two-cylinder, outside-connected express passenger engine, Class E3. (b) Cross-section of standard 100 lb/yd (49·61 kg/m) Stevens–Vignoles rail, as used in the stress measurements. The difference between the broken and unbroken profiles denotes wear. (c) Strain gauge used to measure lateral displacement of the rail. (d) Stresses in the base of the rail under engine 602, when hauling test train over the gauged section of track (e), at 70·5 mph (113 km/h). (e) Plan view of the gauged test section, showing the staggered rail joints, and the direction of travel of the test trains (arrow). (Reprinted from *Engineering (UK)*)

trailing truck; 301,100 lb (1,340 kN) on the six axles of the tender, giving a total weight, engine and tender, in working order of 648,100 lb (2,882 kN). The test train consisted of seven passenger coaches, of four types (98,140; 104,400; 111,200; 113,500 lb; 436; 462·1; 494·5; 504·5 kN) and a caboose or brake van of 38,280 lb (170·15 kN) giving a total load of 766,200 lb (3,409 kN).

These tests confirmed the theory which assumed that rail could be treated as a flexible beam, showing that it was '... well substantiated by numerous experiments in the field ...' though it was discovered that there could be quite large discrepancies between theoretical and actual rail stresses under a particular wheel of a particular engine, the measured to calculated ratio being from about 0·62 for the trailing truck of one of the freight engines to values of 0·7, 0·9, 1·2 for the coupled wheels. It was concluded that factors other than hammer-blow were at work, resulting in actual stress exceeding the theoretical:

> ... The results indicate ... that the steam thrust and inertia components along the connecting rod, and perhaps also the friction of the reciprocating parts, introduce cyclic variations in the main-driving wheel load, so that the resultant stress-revolution relation is no longer a simple sine wave like that calculated. ...[95]

For the freight engines tested, the greatest dynamic augment was under the rear coupled wheels, amounting to 22,100 lb (98·25 kN) at 65 mph (105 km/h), the associated rail stress fluctuating from near zero to 25,000 lb/in² (1·72 × 10⁸ N/m²). The investigations showed that the high-speed express engine, Class E3, worked the track less destructively than the goods engines, and in this case the theoretical and experimental predictions of rail stress were much closer together. Rail stress was described as 'very moderate' for all wheels except the left-hand main driver, and the effects of forces such as steam pressure were less in evidence (Fig. 15d). Lower wheel speed was regarded as playing a significant part in the better performance of the express engine, compared to the freight, the former having coupled wheels of 75 in dia. (1,906 mm) and the latter having coupled wheels of 62 in dia. (1,576 mm). Thus at 80 mph (128·8 km/h), the freight engine ran at 7·1 rps compared to the passenger engine's maximum rotational speed of 6·7 rps at 90 mph (144·9 km/h).

Slight increases in driving wheel diameter for express engines were to be adopted by several companies as a result of such tests, to reduce hammer-blow and the tendency towards wheel lift. The good results with express engine No. 602 were put down to the reduction in weight of the machinery, and the careful rebalancing to which it had been subjected before the tests, prior to working high-speed services like the '400'.

It was concluded that such engines, if properly balanced, could safely operate up to 90 mph (144·9 km/h) without damaging Stevens–Vignoles rail weighing 100 lb/yd (45·36 kg/m) or more. While by no means the heaviest main-line section—some railways used 131 lb/yd (59·4 kg/m) and heavier sections were shortly to be introduced—weights of 100/110 lb/yd

(45·36/49·9 kg/m) were common. This relatively light section was chosen for the tests in order to increase the effects of hammer-blow and also to help determine the lightest rail section suitable for main-line service.

As a general rule, companies like the Chicago and North Western operated steam engines so that the computed rail stress in the centre of the base of the rail did not exceed 30,000 lb/in² (2·067 × 10⁸ N/m²), giving a safety factor of about 2 for the average rail steel elastic limit of 60,000 lb/in² (4·134 × 10⁸ N/m²).

This safety factor covered eight points, listed in the report, other than hammer-blow effects: lateral bending of the rail; eccentric loading; temperature stresses; excess loading of one rail on curves; reduced rail strength by wear; track-bed defects, or sleeper imperfections; flats on wheel tyres and 'out-of-roundness'; and the greater stress, by about 20 per cent, in the rail head compared to the base.

The later trials, reported in the United Kingdom in 1948, followed much the same pattern as the first series, but they involved a greater range of rolling stock which was sent to the same test track by several companies in the summer of 1941. Types tested were generally larger, including 4–6–2, 4–6–4 and 4–8–2. All were two-cylinder, outside-connected types. The same type of rail was used, and measurements were made of lateral forces on the track by mounting a 45 ft (13·72 m) section on roller-bearing tie-plates as shown in Fig. 15c.

The rail rests on a plate, 1, supported by the roller bearings, and a bolt, 2, transmits motion up to $\frac{1}{8}$ in (3·175 mm) to the vertical member, 3, carrying the strain gauge. This measurement of lateral force enabled quantified investigation of the nosing couple on the engine, and the resulting rail force arising from the action of the reciprocating masses:

> ... The forces imposed on the rails fluctuate cyclically with the revolutions of the coupled wheels, and depend on the angular position of the unbalanced force vectors. To relate these with the situation of the strain gauges mounted along the two test lengths of track, slow-motion cinematograph films were taken to record the position of the crank-pin on one side of the locomotive as it passed each test section ...[96]

Furthermore, the accelerations of different parts of the engine and the stresses in different members, such as the wheel mountings, were taken. The findings included the unexpected, though one must be careful about generalizing from one engine tested to another. It was found with the 4–8–2 locomotive, for example, that the stresses in the base of the rail under the two-wheeled trailing truck were of the same order of magnitude as under the coupled wheels, despite the truck wheels being nominally in perfect balance and performing a carrying function only. This was put down to rolling and swaying of the engine, plus the transfer of hammer-blow effects through the spring gear to the rear truck. What is more, for this engine, no definite link could be established between the lateral force and the variation of the nosing couple. The investigation also revealed that the behaviour of the driving wheels on the rail was considerably more complex than had been supposed:

> ... The depression of the wheel and rail, due to ... (the) ... cyclic vertical force causes the coned tread of each driving wheel in turn to make contact with the rail at a different lateral position. Thus each driving wheel runs so that its effective tread diameter is continually increasing and decreasing slightly, causing each driver in turn to run ahead or lag behind the other, and so to give rise to two pressures between each driving-wheel flange and the rail for each revolution ...[97]

The report, however, did not suggest that these factors greatly effected the maximum lateral force recorded by the lateral strain gauges.

The third series of trials, in August 1942, involved some very interesting model experiments on a 4–6–4 engine. It was suspected that the peaks of rail stress, as measured, were out of phase with the maximum vertical force or hammer-blow by about 30 deg. lag. The cause was suspected as being track elasticity, and a scale model was made of the main axle, the sprung and unsprung weights, the revolving and reciprocating masses, and the springing. This rested in a framework capable of vertical motion, supported on a simulated elastic track consisting of rollers on laminated springs. When the model driving axle was revolved at 6–8 rps, corresponding to over 100 mph (160 km/h), high-speed photography of crank position and laminated spring deflection clearly demonstrated the existence of this lag, suggesting that the elasticity of the track was responsible.

The overall conclusions agreed with those of the British investigators, bearing in mind that American design policy favoured two-cylinder engines for the greatest power outputs in express work, and—with one exception—was to reject multi-cylinder drive as a means of reducing the overbalance needed.

Granted that two-cylinder engines needed overbalance, the investigators suggested that it lie between certain limits.

Some American systems had expressed overbalance as a percentage of engine weight, but the investigators pointed out that with respect to the tendency of the driving wheels to rise and fall:

> ... Since the axles are not sprung, and the inertia of the wheel-axle assemblies is relatively small, the hammer-blow, or dynamic augment, is not significantly affected by the mass of the engine, but depends primarily upon the inertia of the reciprocating counterbalance weight in the coupled wheels. The maximum amount that can be tolerated is limited by the additional stress which the rails can safely carry.
>
> Hence the practice recommended by the investigators has been to specify an overbalance force, at the maximum service speed of the locomotive, as great as would ensure that the track would not be damaged by the downward forces, and certainly not so great as

would cause the wheels to leave the rails during the upward part of the cycle fluctuation. . . .[98]

Expressing the overbalance as a force and not a percentage of the weight of the reciprocating parts, as was common in Britain, and remarking that lateral forces and motions of the engine were of minor consideration in deciding it for a particular type, the investigators recommended 100 lb (444·8 N) per coupled wheel as a target figure, with a maximum of 200 lb (889·6 N) per wheel above which hammer-blow became unduly great.

A more specific recommendation was 150 lb (667·2 N) overbalance in the main driving wheels (which carried the connecting rod crank pins); 200 lb (889·6 N) in the other coupled wheels. It was stressed that the reciprocating weight left unbalanced on each side should not exceed 0·4 per cent of the engine weight in running order. These were regarded as maximum permissible values, the optimum being considerably less.

Locomotive designers faced a very real problem if their increasingly heavy two-cylinder engines were to avoid damaging the track because, as the investigators pointed out, when the reciprocating parts were extremely heavy, it might not be possible to comply with the recommendations. For example, after adding the maximum recommended balance weights to the driving and coupled wheels, there might be a reciprocating mass on each side exceeding 0·4 per cent of the engine weight. The total mass of the reciprocating parts was a crucial factor, and it was emphasized that this should be kept as small as was practically possible:

> . . . If, for example, the total reciprocating mass amounts to less than 0·4% of the weight of the engine in working order, then the locomotive will presumably ride satisfactorily with no reciprocating compensation whatever, and hence, if balanced for rotating masses, with no dynamic augment in any wheel and, consequently, no . . . damage to rails and track . . .[99]

Rail Section 1940–50

Between 1940 and 1950, these findings were reflected in locomotive and rail design, though steam engine development declined very sharply after 1948 when it became obvious that other traction systems were destined to become standard.[100] During this last decade of steam traction, there were interesting developments in rail, which had a considerable effect on limiting the size and weight of the engines.

Although the rail-working investigations had been conducted with 100 lb/yd (49·61 kg/m) rail, much heavier section was in use on main lines carrying dense traffic. The weight of rail used was not simply determined by the maximum axle load of the rolling stock, but was also set by traffic volume and speed.

As Dick points out:

> . . . Where traffic ranges from 1 to 8 million gross-tons annually and where train speeds range from 40 to 60 mph, the 100-lb section is

adequate. With traffic ranging from 6 to 15 million gross-tons an-
nually and with train speeds from 50 to 75 mph, the 115-lb section
is justified. The rail size is increased to 132-lb when traffic ranges
from 12 to 25 million gross-tons and with train speeds from 50 to
80 mph, and to the 140-lb and 155-lb sections for traffic ranging
from 20 to 38 million gross-tons and for train speeds of from 50 to
80 mph and higher. These selections are based on the rail handling
both freight and passenger traffic, but, if for freight only, the next
lower rail size can be used . . .[101]

Dick admits that the need to standardize meant that most railways
stocked one or two of the newer rail sections, rather than the full range;
and that the trend to heavier rail was justified by reduction in maintenance labour costs. The average weight for rail on Class 1 railways had
gone up from 82·89 lb/yd (41·1 kg/m) in 1921 to 103·5 lb/yd (51·36 kg/m)
in 1953, when the switch to diesel traction was well under way.

In 1921, 15 per cent of the track was rail weighing over 100 lb/yd
(49·61 kg/m), but in 1953 this was 59 per cent. For 1953, when some of
the long-distance express trains on Class 1 railroads were still steam
hauled, rail section in the 110–119 lb/yd (54·6–59 kg/m) range formed
21 per cent of track mileage; and section of 130–139 lb/yd (64·45–68·96
kg/m) formed 18 per cent. Other weights were laid down in lower percentages of mileage. Very heavy section, such as the Pennsylvania Railroad's 140 lb/yd (69·43 kg/m) PS section and the 155 lb/yd (76·9 kg/m) PS
section were limited to track bearing very heavy traffic on divisions with
much curvature, where rail loading was more severe than on the straight.
Generally, one could not assume that the most powerful and heavy locomotives would be restricted to the heaviest rail sections, especially as
operating policy was to utilize types capable of working over the greatest
possible percentage of the system. This was one reason why the rail-working tests were conducted on 100 lb/yd (49·61 kg/m) rail at a time when
much heavier sections were in use, for even though heavy section was
shown to be economical, considerable lengths of lighter track remained in
service, too recently laid to make relaying with stronger rail worth while.
Certainly, locomotive designers could not expect to escape the problem of
hammer-blow by persuading the railway companies to relay track on a
massive scale with the heaviest sections then available: they had to design
with sections of about 110 lb/yd (54·6 kg/m) in mind, remembering that
80 per cent of the American Class 1 system was 100–120 lb/yd (49·61–
59·5 kg/m), with 59 per cent weighing less than 110 lb/yd (54·61 kg/m).

In the United Kingdom, and to a lesser extent Continental Europe, the
problem was not as acute because locomotive size, weight and power had
not increased to the same extent as in the United States between the early
1920s and the late 1940s.[102] In the United Kingdom during the mid-
1920s, an express passenger locomotive was usually of the order of 90 tons
in running condition without tender. The weight of the last British express
steam engine, a 4-6-2 built in 1954, was 101 tons in running order

without tender.[103] In the United States, the increase in locomotive weight in running order without tender, over the same period, was from approximately 150 tons for a 4-6-2, the usual express engine in the 1920s, to 230 tons for a 4-8-4 constructed in the 1940s.[104] Tender weight increased much more in the United States, reaching 200 tons fully laden by 1945.[105] Furthermore, although the weight increase for the larger freight locomotives (without tender) was not so marked in the United States during this period, by 1945 the speeds were much greater, with 270 ton freight engines of the 4-6-6-4 type running at 70 mph (112·7 km/h) compared to speeds of 20 mph (32·2 km/h) or less for heavy goods locomotives in the 1920s.[106] Locomotive power increased greatly between 1925 and 1945, so that at the latter date American designers were attempting to produce two-cylinder 4-8-4 high-speed types generating about 7,000 ihp (5,220 kW), such as the New York Central System Class Slb.[107] The cylinders of these engines were very large, 635 × 813 mm, and despite the use of light alloy components where possible, the size and weight of reciprocating parts had reached, or was very close to, the maximum.

Greater power, within the form of the two-cylinder 'Stephenson' type, could hardly be achieved without heavier rail section and track bed, which was out of the question for financial reasons. Hence, by 1940, locomotive engineers were exploring alternative designs, capable of the required performance, but which kept down rail stresses.

The Reduction of Rail Stress and Locomotive Design

The problem of keeping down rail stresses was virtually insoluble for the 'Stephenson' locomotive if very high indicated powers were required: say, over 6,000 ihp (4,475 kW) in the United States; over 3,500 ihp (2,610 kW) in France; over 2,500 ihp (1,863 kW) in the United Kingdom. In Europe the problem could be resolved by using multi-cylinder engines, such types, with three or four cylinders, having been used for the greatest power outputs since the nineteenth century.

Although two-cylinder types were preferred whenever possible, it was accepted that for the greatest power outputs three or four cylinders were necessary. Towards the end of steam, three cylinders became the maximum, some such designs running without any overbalance and hence no hammer-blow. France operated compound-expansion engines with three, four and six cylinders, the latter being an experimental engine introduced in 1940, and simple expansion three-cylinder machines, and the final steam designs, never realized in the face of electrification and dieselization, were for compounds using three cylinders for the greatest powers.[108] The final express passenger steam type in the United Kingdom, a 4-6-2 built in 1954, was a simple-expansion, three-cylinder machine.[109] Using three or four cylinders enabled some degree of balance to be achieved between the reciprocating parts themselves, so that overbalance, hammer-blow and, hence, rail stress could be kept low.

Some of the plans for a new generation of steam engines, to be built in

France during the 1950s, giving high cylinder powers of above 4,000 hp (2,982 kW) at speeds of up to 156 mph (250 km/h) showed three cylinders and compound expansion,[110] while a German project, dating from 1941, for a 10,000 ihp (7,460 kW) steam engine capable of reaching 250 km/h indicates two 4–8–4 units, coupled together, each having five cylinders (two high-, three low-pressure).[111] Hence it would seem that using multi-cylinder drive would have been the European solution to the problem had steam traction been developed further. Indeed, there were experiments to develop multi-cylinder drive to produce a high-speed steam motor, geared to the driving wheels, with several cylinders in each unit, with lightweight reciprocating parts, and the pistons and cranks set to give self-balance, thus eliminating overbalance and hammer-blow completely and reducing the tendency to slip. Isolated efforts in this direction were tried in Germany and the United States, but without success. For example, as four cylinders driving onto one axle could be self-balanced, the Baltimore & Ohio Company, probably influenced by the German experiments, produced designs for the 'Constant Torque Locomotive', a 4–8–4 express engine with each driving axle powered through gearing by a four-cylinder high-speed steam motor, giving sixteen cylinders in total. No overbalance was needed and the plan was to omit coupling rods between the driving wheels, resulting in a design that should have exerted no hammer-blow at all.

The engine was never actually built, though the German railways did produce a similar 2–8–2 type powered by four two-cylinder 'V' twin Lentz steam motors, built by Henschel in 1941, and taken to the United States after the war. Although these steam motors did reduce hammer-blow, they were never a success, being unreliable and expensive to maintain.

Another attempt to eliminate hammer-blow, and to gain other advantages, employed electric transmission, with current generated by turbines or high-speed steam engines, but the six major experiments with steam-electric traction on the railways were all total failures as has been described elsewhere.[112]

The 1930s witnessed a revival of the Duplex scheme as designers sought to produce larger and more powerful reciprocating steam engines in which component stresses were minimized and which did not work the permanent way destructively. A preliminary step in this revival was a French design of 1932.[113] This was a four-cylinder compound-expansion freight engine, of the 2–10–2 wheel arrangement, constructed for the P.L.M. company. Contrary to customary French practice for non-articulated engines, all four cylinders were placed externally, the two high-pressure cylinders driving the rear group of six-coupled wheels; the two low-pressure cylinders driving the front group of four-coupled wheels. In this form of Duplex, the two groups of driving wheels were linked by internal coupling rods, which not only greatly reduced the tendency to slip but prevented the two sets of drive from getting out of self-balancing phase. Consequently, though the type was not developed further in France, this particular class functioned successfully enough and had a long life.

In the United States, traction policy involved keeping the engine simple, with no more than two cylinders on a non-articulated chassis, and
complexity in the form of inside cylinders—let alone steam motors—was
decisively rejected. But in the 1930s it was becoming difficult to design
increasingly powerful engines of the traditional two-cylinder type; yet
unless the weight of reciprocating parts was reduced, and hammer-blow
kept down, there seemed no way of increasing performance without relaying main lines with heavier rail.

One solution,[114] suggested by the Baldwin Locomotive Works in 1932,
combined features of the Haswell 'Duplex', the Shaw 'Balanced Compound' and the 'Double-Single', having two groups of driving wheels,
each driven by two external cylinders fastened to a rigid frame, the second
pair of cylinders lying between the two groups of coupled wheels. The
layout was similar to the French engine of the same year, but Baldwin's
proposal omitted linking the two sets of driving wheels through rods, as
this would have necessitated internal cranks which were contrary to American practice.

This proposal for a new kind of Duplex was put to the Baltimore &
Ohio, the Florida East Coast (1935) and the New York–New Haven
(1936) companies, but was rejected on the grounds of the over-long wheelbase intrinsic to the scheme, despite the claim that two sets of four-coupled
driving wheels would corner better than one set of eight-coupled driving
wheels. The idea had some attractive features, for it enabled the weight
of individual reciprocating masses to be greatly reduced, and—by dividing
the drive—reduced the overbalance on any given axle. In 1937 the first
twentieth-century American 'Duplex' appeared, designed by W.B. Whit-
sitt, Assistant Chief of Motive Power and Equipment of the Baltimore &
Ohio Railroad.[115]

As shown (Fig. 16) this engine, No. 5600, employed the Duplex principle by having four externally-placed simple-expansion cylinders, driving
two quite independent groups of coupled wheels, this latter feature making
use of the principle first demonstrated in the 'Double-Singles'. Engine
5600 was unusual in more ways than one. It was the first 4–4–4–4; the
cylinders were cast integral with the frame; the wheelbase was kept to
normal dimensions by placing two of the cylinders behind the coupled
wheels beside the firebox (using a layout employed in the Delaware &
Hudson 4–8–0 four-cylinder triple-expansion locomotive of 1933); and a
water-tube furnace was used. Though capable of hauling 810 ton trains
at 82·5 mph (132 km/h), the Duplex was not developed further by the
Baltimore & Ohio RR.

Engine 5600 did, however, exert considerable influence on other designs
aimed at reducing rail stresses. It impressed the French engineer André
Chapelon, who travelled on it during 1938.[116] At the time, Chapelon was
projecting an experimental six-cylinder compound expansion 2–12–0 locomotive incorporating reheat, steam jacketing and multi-cylinder drive to
twelve-coupled wheels, in part to minimize destructive track working.
Preliminary drawings of this engine, dating from 1937, show a 'Duplex',

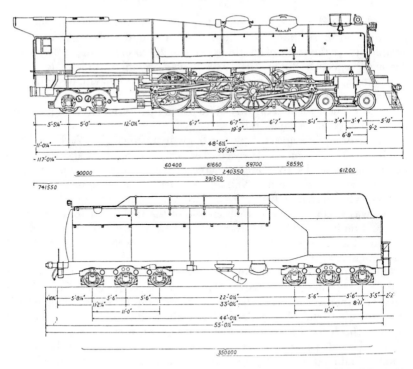

Figure 16.   The first American 'Duplex', No. 5600, Baltimore & Ohio RR, 1937. There were four cylinders 458 mm dia. × 672 mm stroke, all externally placed, driving eight driving wheels 1·92 m dia. grouped in two groups of four, giving a 4-4-4-4 type. The arrangement of cylinders enabled the wheelbase to be shorter than with the Pennsylvania designs (Fig. 17). Engine weight in running order was 175 tonnes; 108 tonnes resting on the driving wheels. The tender weighed 159 tonnes loaded. (Reprinted from *The Locomotive, Carriage & Wagon Review* (London).)

and it seems likely that Chapelon was influenced by the P.L.M. engine of 1932 as well as the American engine with its two sets of independent coupled wheels because the two high-pressure cylinders of the Chapelon design were to drive the rear six-coupled wheels while four low-pressure cylinders would drive the front set of six.[117] When the actual engine emerged for trials, greatly delayed by the Second World War, the 'Duplex' form was abandoned and all twelve wheels were coupled together, probably to minimize slipping. This engine departed very radically from the other designs in as much as four of the six cylinders were internally placed, between the frames, whereas all Duplex engines as built had outside cylinders.

    No. 5600 encouraged the Baltimore & Ohio Company to carry the divided drive principle further and to consider plans for a sixteen-cylinder 4-8-4 express engine, with each driving axle powered by four Besler steam motors, driving through gears.[118] Motion was to be immersed in an oil bath, with cut-off control exercised through electropneumatic circuits. The four sets of two driving wheels were to be uncoupled, resulting in a 'Quadruple-Single' type, and it was predicted that no balancing of any kind would be necessary.

It was further expected that the independence of each wheel set would facilitate good cornering without excessive tyre wear. This project, the 'Constant Torque' locomotive, was never realized, though a similar German design was constructed with external motors.

The projected combination of oil bath and lack of coupling rods would probably have led to excessive and destructive wheel slip.

The French constructed a steam-motor locomotive, with eighteen cylinders and a water-tube boiler, which proved a failure.[119]

Though not repeated by the Baltimore & Ohio Company which built it, No. 5600 was deemed successful enough to encourage the Pennsylvania Railroad to start applying the Duplex principle to their most modern motive power after 1939. This was the only major application of the Duplex principle and several dozen very large engines were built.

In the late 1930s, the Pennsylvania Company were seeking to reduce rail stresses while trying to produce a locomotive capable of hauling 1,000-ton trains at 100 mph (160 km/h), the specification suggested by the American Association of Railroads as being necessary for passenger motive power in the 1940s.[120]

The Pennsylvania Company was to carry out two major experiments to meet this specification, one with reciprocating engines (the S1 project) (Fig. 17), the other with a geared turbine (the S2 project) (Fig. 18).

Figure 17. Pennsylvania RR Class S1, No. 6100, four-cylinder simple-expansion 6-4-4-6 'Duplex' type, constructed 1939 to haul 1,000 tons at 160 km/h. All four cylinders were externally placed, driving two groups of four driving wheels. Cylinders were 559 mm dia. × 660 mm stroke. Driving wheels 2·13 m dia. Weight on driving wheels 128 tons. Weight of engine (without tender) in running order 273 tons. (Reprinted with permission of Pennsylvania RR (Now part of Penn. Central).)

Figure 18. Pennsylvania RR Class S2, No. 6200 6-8-6 turbine-mechanical engine, constructed in 1944 to haul 1,000 tons at 160 km/h. The main turbine is over the second and third coupled wheels. (Reprinted with permission of Pennsylvania RR (later Penn. Central).)

The first trial, under the auspices of the American Association of Railroads, resulted in Engine 6100, Class S1 built in 1939 and exhibited at the New York World Fair.[121] It weighed 273 tons in running order without tender, and proved able to haul 1,340 tons at 125 km/h or 3,175 tons at 96 km/h. In this design, inside cranks were avoided by placing all four cylinders outside the frames, the second set lying between the two groups of four-coupled wheels, resulting in a Duplex type similar to the scheme first suggested by the Baldwin Locomotive Company in 1932. Because of this cylinder disposition, No. 6100 had a much longer coupled wheelbase than the earlier No. 5600. As no Duplex used articulation—because the Mallet or Beyer–Garrett types were not thought suitable for very high speed—these rigid wheelbases were intrinsic to the type and were to prove an undesirable feature. Unlike the similar French design of 1932, neither of these prototype American Duplex locomotives had the two groups of four driving wheels coupled together, and this was to prove a design weakness, leading to slipping.

The basic idea of the Duplex was that using four cylinders reduced the maximum weight of any single reciprocating component and that some degree of self-balance could be obtained. Furthermore, there were two main driving axles, so the balance weight fastened to the main drivers was less than if there were only one, as for the orthodox two-cylinder engine of equal power. Large-diameter wheels were used to keep the rotational speed low at high speed, thus countering the tendency towards wheel lift, and it was argued that two independent groups of four-coupled wheels passed round severe curves more easily than eight coupled together. Generally, it was hoped that the Duplex would give very high performances while reducing hammer-blow and rail stresses. Certainly, the Pennsylvania Duplex S1 was very fast and powerful: a 6-4-4-6 type, it was capable of sustaining over 6,000 ihp (4,475 kW) indefinitely, with maxima of over 8,000 ihp (5,964 kW), and could attain speeds in excess of 100 mph (160 km/h). It was followed by similar, but smaller, Duplex locomotives for passenger and freight, built after 1941 with wheel arrangements 4-4-4-4, 4-4-6-4 and 4-6-4-4.[122] All were failures, inflicting on the track the damage they were meant to reduce. They were not articulated engines, and the rigid wheelbase, much longer than with an ordinary two-cylinder 4-8-4 locomotive, caused wear on curves.

The greatest defect was slipping, most likely due to the coupling rods being omitted between the two groups of driving wheels. Just as a single-driver engine was far more prone to wheel slip than a four-coupled engine, as shown in Fig. 4, so two groups of four driving wheels slipped more readily than a single group of eight, all coupled. The slipping was very destructive of track (Figs. 19 and 20), though once started it was probably less so with a Duplex than an orthodox 4-8-4 because of relatively low overbalance.

Because the two groups of driving wheels could slip independently of each other, the two sets of driving cranks could get out of the phase in which the four pistons best balanced each other, and though the Duplex engines were powerful and very fast, they are generally regarded as examples of design failure. Indeed, the only successful 'Duplex' class was the above-mentioned series of four-cylinder compound 2-10-2 freight engines used in France, which significantly had the two groups of driving wheels, of four and six, linked by internal coupling rods, preventing them from getting out of self-balancing phase and reducing the tendency to slip.

The other Pennsylvania experiment, of 1944, was to fit a geared turbine to an engine basically similar to the S1.[123] The turbine locomotive, No. 6200, Class S2, was a 6-8-6 rated at 6,900 hp (5,145 kW) turbine power, weighing 253 tons in running order without tender. It was quite free from hammer-blow because there were no reciprocating parts; only revolving masses had to be balanced and this could be done completely by driving-wheel weights. Though the S2 worked track less destructively than reciprocating steam engines, it had—like a similar, though much smaller turbine engine in the United Kingdom[124]—a high steam consumption at low speed, and neither was deemed promising enough to continue. All turbine experiments proved relative failures, largely due to their inefficiency over a wide speed range; their inability to reverse without secondary turbines; their need of a geared transmission or electric drive; and the lack of thermodynamic advantage unless a vacuum condenser was fitted. Engine No. 6200 was a turbine-mechanical unit, exhausting to atmosphere; a second, smaller turbine was carried for reversing.

The rail-stress problem was not solved, as far as United States practice was concerned, and it is difficult to see how greater powers and speeds could have been supplied using a two-cylinder 'Stephenson' locomotive without demanding a much heavier rail and track structure. The fruitful union of 'Stephenson' locomotive and 'Stephenson' track structure, dating from the Liverpool and Manchester Railway of the 1830s, was disrupted by the imminent obsolescence of the steam traction system, signalled by the inability of locomotive engineers to solve such problems as were related to rail stress. Systems harmony was at an end: the Machine-Ensemble was

Figure 19. Stevens-Vignoles rail damaged by locomotive wheel slip. (Reprinted with permission of Sperry Rail Services.)

breaking down. There were many other reasons why steam traction was obsolete, or almost so, but the impending impossibility of designing types capable of much greater power and speed, within the two-cylinder 'Stephenson' form, was recognized in the United States around 1945 (Fig. 21). With all attempts to produce a new type of engine,[125] such as the

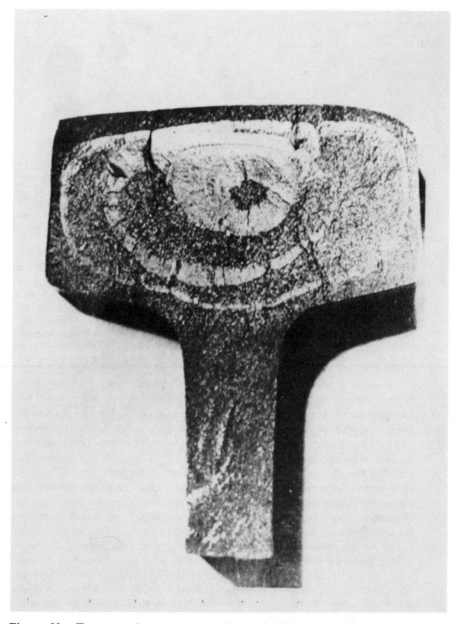

Figure 20. Transverse fissure, caused by impact loading, spreading within the head of a Stevens–Vignoles rail. The growth of the fissure in stages is represented by a series of concentric rings inside the head. (Reprinted with permission of Sperry Rail Services.)

Figure 21. The final stage of development of the Stephenson two-cylinder, simple-expansion steam engine. Outside-connected 4-8-4, type S1b, built 1945 for New York Central System. Two cylinders, 635 mm dia. × 813 mm stroke, drove eight coupled wheels of 2 m dia. Engine weight fully loaded (without tender) 214 tonnes; weight on driving wheels 125 tonnes. Note double-disk wheel construction with tubular stays and crescent-shaped driving wheel balance masses. (Reprinted with permission of New York Central System (now part of Penn. Central).)

steam-electric, meeting with failure, the railways accepted superior modes of traction which were free from reciprocating masses altogether. These newer traction systems, soon to replace steam in all but the technically backward nations, brought problems of their own such as the tendency of early diesels to slip because of high starting tractive effort. This could lead to 'rail burning', but slip control applied to diesel and electric locomotives reduced this. Slip eliminators had been tried on some steam engines, but they were crude, mechanical devices, not as effective as the electric and electronic devices used in modern traction systems.

Another type of track damage was caused by the heavy unsprung masses of electric traction motors increasing impact loading on rail, which is still an important factor in wear though it has been reduced by better motor suspension design. Despite these problems with the newer traction systems, the basic 'Stephenson' form of track structure, using 'Stevens–Vignoles' rail on transverse sleepers, seems capable of indefinite development using rail sections of weights much less than would be needed if steam traction were employed.

Conclusion

The relationship between Stephenson steam locomotive and Stephenson track structure affords a case study in the useful union between two technologies evolving towards the point where one must be abandoned in favour of the other. Several major phases of development can be identified, each being capable of subdivision.

The first saw the emergence of the basic steam locomotive and track structure on the Liverpool and Manchester Railway, under the guidance of George Stephenson, aided by his son Robert, and their colleague, Locke. The next phase witnessed the recognition that the universally-employed two-cylinder engines were destructively working the permanent way in a fashion that could be reduced by techniques like balancing. There then followed a long period of gradual improvement to engine and track, which saw the use of steel for tyres and rail, and the growth of balancing theory to describe the forces exerted on the rail by the action of the weights.

Generally, it was possible, during this phase, to solve problems relating to engine and track without departing from the Stephenson forms. This long spell of improvement lasted well into the twentieth century until the performance of steam engines, greatly enhanced by advances in design, threatened the unity of the Stephenson track-engine system through excessively destructive rail working. The twentieth century witnessed a great development in techniques for investigating all problems related to rail working, fracture and other forms of failure, some of which could be solved by reformed methods of manufacture, though others—such as were associated with excessive dynamic augment, wheel bounce and slip—demanded redesign of either the Stephenson track or the Stephenson locomotive. The harmony between the two was coming to an end.

Since the late nineteenth century it had been necessary to develop scientific techniques for balancing locomotives if rail damage was to be reduced and the usefulness of the Stephenson machine-ensemble preserved, but by the 1940s it was becoming increasingly difficult to avoid impasse. The final phase encompassed the end of steam traction, all efforts to redesign it to meet modern needs failing to such a degree as to signal its obsolescence beyond doubt.

It was no longer possible to solve the many design problems, including those associated with the need to reduce rail stresses, without extensive and prohibitively expensive reconstruction of permanent way. However, the abandonment of steam traction enabled the Stephenson track structure to be retained in service with newer traction systems. Both have evolved in harmony, and as yet there is no sign that the point where one needs to be drastically redesigned or abandoned is approaching.

Some two hundred years after the birth of George Stephenson in 1781, there is every reason for regarding the universal form of railway track as being as much his memorial as the once standard form of steam locomotive.

Notes and References

1. The basic machinery of the Stephenson locomotive is described in J.T. Hodgson and C.S. Lake, *Locomotive Management*, Tothill Press, London, 1954, esp. pp. 123–222.

2. G.V. Lomonossoff, *Introduction to Railway Mechanics*, Milford-Oxford University Press, 1933.

3. W. Schivelbusch, *The Railway Journey*, Basil Blackwell, Oxford, 1980, pp. 19–24.

4. M.C. Duffy, 'Technomorphology and the Stephenson Traction System', *Transactions of the Newcomen Society*, Vol. 54, 1982–3, pp. 55–78.

5. H. Greenleaf and G. Tyers, *The Permanent Way*, Winchester Publ. Ltd., London, 1938; J. Elfreth Watkins, *The Development of American Rail & Track*, US House of Reps., Annual Report Board of Smithsonian Inst., 1889, Misc. Document 224, Pt. 2, 51st Cong., 1st Session; Von Oeynhausen and Von Dechen, *Railways in England*, Newcomen Society Extra Publ., Heffer, Cambridge, 1971, reprint of Engl. trans. of German reports, 1826, 1827.

6. Anon, *A Century of Locomotive Building 1823–1923*, David & Charles, Newton Abbot, 1970, reprint of Reid Publ. 1923 first edn; E.L. Ahrons, *The British Steam Railway Locomotive 1825–1925*, Ian Allan, 1966 reprint of 1927 first edn; W.A. Tuplin, *The Steam Locomotive*, Adams & Dart, Jupiter Books, 1974.

7. M.C. Duffy, 'George Stephenson and the Introduction of Rolled Railway Rail', *International Journal of Mechanical Working Technology*, Vol. 5, 1981, pp. 309–42.

8. Z. Colburn, *Locomotive Engineering & the Mechanism of Railways*, 2 vols., William Collins, Sons & Co., London and Glasgow, 1871.

9. A. Chapelon, *La Locomotive a Vapeur*, Balliere et Fils, Paris, 1938, 1952.

10. Ahrons, as note 6; O.S. Nock, *The British Steam Railway Locomotive, 1925–1965*, Ian Allan, 1969.

11. W.T. Tomlinson, *North-Eastern Railway*, David & Charles, Newton Abbot, 1967, reprint of 1914 original. Contains account of Stockton & Darlington Railway and failure of suspension bridge over Tees, p. 187.

12. M.J.T. Lewis, *Early Wooden Railways*, Routledge and Kegan Paul, 1970; B. Baxter, *Stone Blocks & Iron Rails*, David & Charles, Newton Abbot, 1966, esp. pp. 37–58; B. Morgan, *Civil Engineering: Railways*, Longmans, pp. 10–23; J. Simmons, *The Railways in England and Wales 1830–1914*, Vol. 1, Leicester Univ. Press, 1978, pp. 148–50; Z. Colburn and A.L. Holly, *The Permanent Way*, New York, 1858; E.E. Russel Tratman, *Railway Track & Track Work*, 3rd edn, McGraw-Hill, New York, 1909.

13. C. Hadfield and A.W. Skempton, *William Jessop, Engineer*, David & Charles, Newton Abbot, 1979, pp. 168–83.

14. Tomlinson, as note 11, p. 77, pp. 89–90; Morgan, as note 12; Baxter, as note 12.

15. T.J. Donaghy, *Liverpool & Manchester Railway Operations 1831–1845*, David & Charles, Newton Abbot, 1972, pp. 32–6; Baxter, as note 12, p. 55; F. Whishaw, *Railways of Great Britain and Ireland*, David & Charles, Newton Abbot, 1969, reprint of 1842 edn contains review of early rail sections. Alternative track forms are reviewed by W. Bridges-Adams, 'Permanent Way', *Proc. I.C.E.*, Vol. XI, 1851-2, pp. 244–98; Vol. XVI, 1856-7, pp. 226–98; F. Fox, 'Iron Permanent Way', *Proc. I.C.E.*, Vol. XX, 1860-1, pp. 259–91.

16. H.C. Archdeacon and M.H. Dick (eds), *The Track Cyclopedia*, Simmons-Boardman, Omaha, 1978; C.L. Heeler (ed.), *British Railway Track*, Permanent Way Institution, 1979. Reviews of modern permanent way practice.

17. Heeler, as note 16, esp. pp. 57–61.

18. Watkins, as note 5; Tratman, as note 12; Z. Colburn and A.L. Holly, *The Permanent Way*, New York, 1858.

19. C.P. Sandberg, 'Manufacture and Wear of Rails', *Proc. I.C.E.*, Vol. XXVII, 1867-8, pp. 320–408; J. Timmis Smith, 'On Bessemer Steel Rails', *Proc. I.C.E.*, Vol. XLII, 1874-5, Pt. IV, pp. 69–127; W. Vamplew, 'Railways and the Iron Industry', contained in M.C. Reed, *Railways in Victorian Economy*, David & Charles, Newton Abbot, 1969, pp. 33–75.

20. As notes 6, 8 and 9; W.W. Mason, 'Trevithick's First Rail Locomotive', *Trans. Newcomen Society*, April 1932; H.W. Dickinson, *A Short History of the Steam Engine*, Univ. Press, Cambridge, 1938; R. Young, *Timothy Hackworth and the Locomotive*, Scolar Press, Ilkley, 1975, reprint of 1923 first edn.

21. As note 6; R.S. Hartenburg, *Kinematics of the Earliest Locomotives*, Proc. World Congress on Theory of Machines & Mechanisms (4th), Univ. Newcastle on Tyne, 8–12 Sept. 1975, Mech. Eng. Publ., 1975.

22. F. Pettigrew, 'Furness Railway Locomotives', *Proc. Inst. Mech. Eng.*, Pt. 3, 1901, Fig. 20, Plate 155. This depicts the very latest express engine of that company.

23. W.A. Lucas (ed.), *100 Years of Steam Locomotives*, Simmons-Boardman, New York, 1957. American practice outlined in J.H. White, *American Locomotives, an Engineering History 1823–1923*, Johns Hopkins, Baltimore, USA, 1968; E.P. Alexander, *Iron Horses 1829–1900*, Bonanza, New York, 1941. This is largely a collection of side elevations and manufacturers' prints; A. Sinclair, *Development of the Locomotive Engine*, Angus Sinclair Co., New York, 1907.

24. W.E. Dalby, 'The Balancing of Locomotives', *Proc. I.Mech.E.*, Nov. 1901, pp. 1157–1208.

25. *Ibid.*, pp. 1162–3.

26. Extreme examples of overbalance and underbalance quoted by H.N. Colam and J.D. Watson, 'Hammer-Blow in Locomotives: can it not be abolished altogether?', *Proc. I.Mech.E.*, Vol. 146, 1941, p. 163.

27. E.S. Cox, 'Balancing of Locomotive Reciprocating Parts', *Proc. I.Mech. E.*, Vol. 146, 1941, p. 148. It would seem that the practice took at least 10–15 years to spread and become common. The side views of US locomotives in Alexander, as note 23, reveal engines fitted with balance weights dating from 1851 (p. 84), 1853 (p. 90), 1854 (p. 96) and 1856 (p. 108) after which it becomes almost universal. The plates show engines without driving wheel balance weights dated 1857 (p. 120), 1858 (p. 124), 1859 (p. 126), and even 1867 (p. 144) and 1873 (p. 166) though it is possible that these later locomotives were balanced by the owning company upon receipt from the makers.

28. Cox, as note 27, p. 148.

29. Dalby, as note 24, pp. 1188, 1202.

30. *Ibid.*, pp. 1190, 1201. A twentieth-century treatment, relating wheel lift to slipping and balancing; and including balancing theory for different kinds of valve gear, crank-settings and cylinder layout is D. Laugharne-Thornton, *Mechanics Applied to Vibrations and Balancing*, Chapman & Hall, London, 1939, pp. 86–116.

31. Dalby, as note 24. Comment by F.V. Russell, p. 1196.

32. *Ibid.*, p. 1202. Communication by S. Rendell.

33. *Ibid.*, p. 1192. Comment by A. McDonnell.

34. *Ibid.*, p. 1202 (S. Rendell). *Ref. to Trans. A.S.M.E.*, Vol. XVI, Dec. 1894.

35. *Ibid.*, p. 1194. Comment by Mr Druitt Halpin.

36. *Ibid.*, p. 1174.

37. M.H. Dick (ed.), *Railway Track & Structures Cyclopedia*, Simmons-Board-man, New York, 1955, pp. 280, 1178-9; M.C. Duffy, 'The Standard Rail Section, Transverse Fissures and Reformed Mill Practice, 1911-1955', *Int. J. Mech. Working Tech.*, Vol. 4, 1980, pp. 285-305; H.F. Moore, 'Rails Investigation', *Metal Progress*, June 1948, pp. 828-31; 'Internal Fissures in Railroad Rails', *Metal Progress (USA)*, Nov. 1935, pp. 46-52.

38. These types are included with a general review of experimental steam locomotives in W. Stoffels, *Lokomotivbau u. Dampftechnik*, Birkhauser, 1976.

39. Ahrons, as note 10, p. 61.

40. Bodmer's system was applied to stationary and marine engines of up to 600 hp (447·6 kW). J.G. Bodmer, 'The Advantages of Working Stationary and Marine Engines Expansively on the Double Crank System', *Proc. Inst. Civil Eng.*, Vol. IV, 1845, pp. 372-399. In these applications, a common arrangement was to link the piston nearest the main crankshaft to the crank, using a connecting rod in the usual manner, but to take the drive from the oppositely moving piston through secondary connecting rods, crossheads and slides, externally placed. This is clearly illustrated in a model, dated about 1870, in the Newcastle Museum of Science & Engineering. On a locomotive there was no room for this system, and the co-axial piston rods and double-crosshead arrangement was used instead.

41. H.T. Walker, 'The Origin of the Balanced Locomotive', *Locomotive, Railway Carriage and Wagon Review* (UK), 14 Feb. 1931, p. 43.

42. J.N. Westwood, *Locomotive Designers in the Age of Steam*, Fairleigh Dickinson University Press, 1978, p. 219.

43. The Great Western type of four-cylinder engines is described by Ahrons, as note 6, p. 332. An example of Dutch practice, with all four cylinders driving the leading axle, is Engine 3737 preserved at the Railway Museum, Utrecht. P. Ransome-Wallis, *Preserved Steam Locomotives of Western Europe*, Vol. 2, Ian Allan, London, 1971, p. 53.

44. O. Reynolds, 'On the Fundamental Limits to Speed', originally printed in *The Engineer*, 28 Oct., 18 Nov., 9 Dec. 1881, reprinted in *Papers on Mechanical and Physical Subjects*, Vol. II, 1881-1900, Cambridge University Press, 1901, pp. 1-24.

45. Ahrons, as note 6, pp. 291, 301, 303. The Pennsylvania engine was of British origin. The question of slipping with four-coupled engines is mentioned by E. Mason, *The Lancashire & Yorkshire Railway in the 20th Century*, Ian Allan, London, 1975 edn (1954 original), pp. 67-8. The unusual phenomenon of slipping induced by shutting the steam off is discussed and related to excessive overbalance.

46. G. Vuillet, *Railway Reminiscences on Three Continents*, Nelson, 1969.

47. O.S. Nock, *The British Steam Railway Locomotive, 1925-1965*, Ian Allan, London, 1966; L.M. Vilain, *La Locomotive a Vapeur et les Grandes Vitesses*, Vincent, Paris, 1972.

48. See references to rail working throughout G.V. Lomonossoff, *Introduction to Railway Mechanics*, Oxford, 1933.

49. Duffy, as note 37, pp. 285-305.

50. R.N. Arnold, 'Impact Stresses in a Freely Supported Beam', *Proc. I. Mech.E.*, Vol. 137, 1937, pp. 217-81, esp. p. 217.

51. *Ibid.*, p. 225.

52. S. Timoshenko, 'Method of Analysis of Statical and Dynamical Stresses in Rail', *Proc. 2nd Int. Cong. Appl. Mech.*, 1926, p. 407; S. Timoshenko and B.F. Langer, 'Stresses in Railroad Track', *Trans. A.S.M.E.*, Vol. 54, 1932, APM-54-26, p. 277.

53. Arnold, as note 50, p. 222.

54. Z. Tuzi and M. Nishida, 'Photoelastic Study of Stresses due to Impact', *Phil. Mag.*, 7, Vol. 21, 1936, p. 448.

55. *Report of the Bridge Stress Committee*, Dept. of Sci. & Industrial Res., 1928, HMSO.

56. H.L. Mason, *Trans. A.S.M.E.*, Vol. 38, 1936, p. A-55.

57. S. Timoshenko, *Zeit. Math. u. Phys.*, Vol. 62, 1913, p. 198.

58. Arnold, as note 50, p. 220.

59. St Venant *et al.*, contributions reviewed in note 48.

60. Arnold, as note 50, pp. 250-2.

61. *Ibid.*, p. 227.

62. *Ibid.*, pp. 235-6.

63. Cox, as note 27, pp. 148-62, esp. Table 1, p. 150.

64. Arnold, as note 50, p. 239.

65. H. F. Moore (ed.), 'Internal Fissures in Railroad Rails', *Metal Progress (USA)*, Nov. 1935, pp. 46-52; 'Rails Investigation', *Metal Progress (USA)*, June 1948, pp. 828-31; Duffy, as note 49.

66. Discussion in Arnold, as note 50, pp. 266-9.

67. J. Simmons, *The Railways in England and Wales 1830-1914*, Vol. 1, Leicester Univ. Press, 1978, p. 146.

68. H.F. Moore and H.R. Thomas, discussion on Arnold, as note 50, p. 268.

69. Cox, as note 27; T.V. Buckwalter and O.J. Horger, *Railway Mechanical Engineer*, USA, Vol. 113, 1939, pp. 95, 132, 186.

70. *Ibid.* See also W.E. Dalby, *The Balancing of Locomotives*, Edward Arnold & Co. Ltd., London, 1930; D. Laugharne-Thornton, *Mechanics Applied to Vibrations and Balancing*, Chapman & Hall, London, 1939, pp. 86-116; Cox, as note 27, p. 148.

71. Cox, as note 27.

72. Representative locomotive types dating from this period are surveyed in P. Ransome-Wallis, *The Concise Encyclopedia of World Railway Locomotives*, Hutchinson, 1959, pp. 319-84.

73. Cox, as note 27, p. 151; these locomotives described by Nock, as note 47, pp. 90, 139-44.

74. Cox, as note 27, p. 152.

75. *Ibid.*, p. 153.

76. Nock, as note 47, pp. 90, 139-44.

77. Arnold, as note 50, p. 227.

78. Cox, as note 27, p. 153.

79. For example, the Southern Railway (UK) Class MN, described by the designer O.V.S. Bulleid, 'Locomotives I have known', *Proc. I.Mech.E.*, Vol. 152, 1945, No. 4, pp. 341-52, esp. 350-2.

80. Cox, as note 27, pp. 150, 153.

81. Colam and Watson, as note 26, pp. 163-6.

82. *Ibid.*, p. 163.

83. *Ibid.*

84. *Ibid.*

85. *Ibid.*, p. 166.

86. Anon, *Locomotives of the L.N.E.R., Part 6B: Tender Engines, Classes O1 to P2*, Railway Correspondence and Travel Society, 1983, p. 124.

87. Anon, 'Determining Counterbalancing Effects by Rail-Stress Measurements', *Engineering (UK)*, 1942, 16 Oct., pp. 301-4; 30 Oct., pp. 341-4; 13 Nov., pp. 381-3; 27 Nov., pp. 420-2; This paper refers to the following primary sources: Anon, *Determining Counterbalancing Effects of C and N.W. Railway Locomotives by*

Rail Stress Measurements, Assoc. of American Railroads, Operations and Maintenance Dept., Chicago, Ill., USA, 1941; 'Report of Special Committee on Stresses in Railroad Track', *Proc. Amer. Railway Engineering Assoc.*, Vol. 253, p. 278 *et seq.*

88. Anon, 'Counterbalance Tests on American High-Speed Locomotives', *Engineering (UK)*, 1948, 13 Aug., pp. 148-50; 20 Aug., pp. 172-4; 15 Oct., pp. 361-2; 29 Oct., pp. 409-11; 5 Nov., pp. 437-9. This paper refers to the following primary sources: Lawford H. Fry, 'Locomotive Counterbalancing', *Proc. A.S.M.E.*, Dec. 1933; *Rly. Mech. Eng. (USA)*, 1934, Feb. and March; H.R. Clarke and K. Cartwright, 'Relation of Locomotive Design to Rail Maintenance', *The Rly. Age (USA)*, 25 March 1939; Anon, *Counterbalance Tests of Locomotives for High Speed Service* and *Supplemental Track Tests of Twelve Locomotives*, Assoc. American Railroads, Chicago, Ill., USA; 'A.A.R. Manual of Counterbalancing for Reciprocating Steam Locomotives', July 1945, Assoc. Am. RR, Chicago, Ill.

89. Anon, 'Determining Counterbalancing Effects by Rail-Stress Measurements', as note 87, p. 301.

90. Anon, 'Counterbalance Tests on American High-Speed Locomotives', as note 88, p. 149.

91. *Ibid.*, p. 148.

92. Anon, 'Determining Counterbalancing Effects by Rail-Stress Measurements', as note 87, pp. 301-2.

93. Anon, 'Counterbalance Tests on American High-Speed Locomotives', as note 88, p. 149.

94. Anon, 'Determining Counterbalancing Effects by Rail-Stress Measurements', as note 87, p. 381.

95. *Ibid.*, p. 383.

96. Anon, 'Counterbalance Tests on American High-Speed Locomotives', as note 88, p. 149.

97. *Ibid.*, p. 174.

98. *Ibid.*, p. 410.

99. *Ibid.*, p. 438.

100. R.C. Bond, 'Commentary on Change from Steam Traction', *Proc. I. Mech.E.*, Vol. 178, Pt. 1, pp. 1-26; P.W. Kiefer, *A Practical Evaluation of Railroad Locomotive Power*, Simmons Boardman, New York, 1945.

101. M.H. Dick (ed.), *Railway Track & Structures Cyclopedia*, Simmons Boardman, New York, 1955, p. 270, p. 274.

102. As notes 47, 72; W.A. Lucas, *100 Years of Steam Locomotives*, Simmons Boardman, New York, 1957.

103. Anon, *British Locomotive Types*, The Railway Publishing Co., 6th edn, 1946; weight diagrams of major types then in use. M. Evans, *Pacific Steam*, Percival Marshall, 1961, esp. pp. 75-9 (final 4-6-2 types); E.S. Cox, *British Standard Steam Locomotives*, Ian Allan, London, 1973 (final British steam types).

104. The typical American 4-6-2, represented by the K4 Class of the Pennsylvania RR, is described in *Engineering (UK)*, Vol. CI, 1916, pp. 405-6; Vol. CII, pp. 98-9, 146-50, 195-6, 244, 295, 350. Final US steam types reviewed by Ransome-Wallis, as note 72, pp. 319-34.

105. For example, the 16-wheel tender of the A.T. & S.F. Class 3765 4-8-4 engine weighed 458,055 lb (2,037 kN) loaded. See Ransome-Wallis, as note 72, pp. 322, 325.

106. A.E. Durrant, *The Mallet Locomotive*, David & Charles, Newton Abbot, 1974.

107. *Engineer (UK)*, Vol. 182, pp. 73-5.

108. Col. H.C.B. Rogers, *Chapelon*, Ian Allan, London, 1972; A. Chapelon, *La Locomotive à Vapeur*, Balliere et Fils, Paris, 1952.

109. Evans and Cox, as note 103.

110. As notes 108 and 47 (Vilain); Chapelon, as notes 9, 108.

111. Vilain, as note 47, pp. 148-51; Gunther, 'Locomotives for High Speeds' (in German), *Glasers Annalen*, Juli 1943, Heft 13/14, pp. 210 *et seq.*

112. W. Stoffels, *Lokomotivbau u. Dampftechnik*, Birkhauser, 1976.

113. The P.L.M. 2-4-6-2 engine was exhibited at the Brussels International Exhibition. Stated to be capable of 2,240 kW at 75 km/h. Note in *Engineering*, 23 Aug. 1935, p. 199 (wrongly describes engine as articulated).

114. Rogers, as note 108, p. 113.

115. Anon, 'Baltimore & Ohio Railroad: Four-Cylinder 4-4-4-4 Type Loco', *The Locomotive*, 14 Aug. 1937, pp. 260-1.

116. Rogers, as note 108, p. 114.

117. Anon, 'Recent Developments in French Steam Locomotives', *The Locomotive*, 14 Aug. 1937, pp. 238-42. Reviews also several designs with 12, 18 cylinders; also turbine driven. Chapelon 'Duplex' 2-6-6-0, p. 239.

118. Anon, 'Sixteen-Cyl. 4-8-4 Loco., Baltimore & Ohio R.R.', *The Locomotive*, Aug. 1937, pp. 311-12. A railcar on the N.Y., N.H. and H.RR. (1937) powered by Besler motors giving 750 kW supplied by steam at 105 bar from a flash boiler is described briefly in Benson and Rayman, *Experimental Flash Steam*, M.A.P. Tech. Publ., 1973, pp. 25-7.

119. As note 117, pp. 241-2. Mention is here made of a plan by Bugatti for a 64-cylinder train for the P.L.M. powered by steam motors.

120. Anon, 'Progress in Railway Mechanical Engineering', *Mech. Eng. U.S.A.*, Vol. LXI, 1939, pp. 863-75.

121. *Ibid.*; *The Engineer (UK)*, 21 July 1939, pp. 62-3, 213.

122. See series by E.C. Poultney, *Engineer (UK)*, Vol. 182, 1946, pp. 458, 460; Vol. 185, 1948, pp. 512-14, 536-7, 562-5. B. Reed, *Pennsylvania Duplex*, Locomotive Profiles, Vol. 2, Profile Publ., 1972.

123. Anon, 'Why a Geared Turbine Steam Locomotive?', *Westinghouse Engineer*, NA5, March 1945, pp. 34-40; *Engineer (UK)*, April 1938, p. 481.

124. Nock, as note 47, pp. 112-17.

125. Stoffels, as note 112. Stoffels describes one of the very last attempts to produce a steam engine which reduced rail stress, the 0-6-6-0 type designed by Bulleid for the Southern Railway (UK) and completed in 1949. The engine was mounted on two six-wheeled bogies, each powered by a three-cylinder motor. Axle coupling was through chains. A very similar engine was developed by Bulleid in the 1950s for the Irish State Railways. Neither type met with success.

Giovanni Francesco Sitoni, an Hydraulic Engineer of the Renaissance

JOSÉ A. GARCÍA-DIEGO

Introduction

Giovanni Francesco Sitoni is a practically unknown figure and this is the first monograph which has been written about a man who was, in his day, an important engineer in Italy and Spain. The leading people of his time were satisfied with the way he carried out the undertakings with which they entrusted him, even though sometimes the results were doubtful or gave rise to controversy. In addition, in his closing years, he wrote one of the oldest known treatises on hydraulic work. All of this being so, one may ask why no one has made a study of him. In my opinion, there are several reasons.

The first reason is of a general nature and will certainly be quite familiar to readers of this journal. The history of technology only came into being more or less recently. Many scholars seek out the slightest details about politicians and writers, for instance, because they suppose that these well-known figures exercised a major influence on the changing course of history and culture. Nowadays, however, this type of historian is giving way to another which believes only in the primary importance of economic factors, often without considering their interdependence on technical developments. Of course, a great deal of the responsibility for this lies with universities and research centres which (though varying greatly in different countries) do not yet supply the necessary funds to finance the quite novel type of research which studying the history of technology demands.

In the particular case of Sitoni, the Italian Renaissance produced a great number of brilliant men in different fields of activity so that studying them, or even only mentioning them, is more difficult than when working on other periods. Nietzsche's pronouncement that 'The European of today is vastly inferior to the European of the Renaissance' is certainly quite unacceptable but still, as with so many of his sayings, there is a certain foundation of truth in it. And one of the outstanding abilities of the Italians was in hydraulic theory and practice. Thus it is not surprising that some figures who may be considered as of secondary importance have been passed over.

In Spain, partly for reasons which I shall explain briefly later, the fine arts and especially architecture flourished in the period covering Sitoni's lifetime. There were, moreover, great engineers such as Juan de Herrera, who was one of those who excelled in both these activities. I shall return to him later. But there were not so many distinguished engineers as there were in Italy. Italians were sent for to work in Spain when they were needed. Their contribution was important and, in the majority of cases, positive. The preference for Italians is explained by the fact that several of the Italian states were under Spanish rule. At the same time, though France was nearer, Spain and France were politically and militarily opposed for long periods. Flemish and German engineers also went to Spain but in far fewer numbers, partly perhaps because of difficulties with the languages.

Before beginning my account of Sitoni's life and work, I must beg leave to deal, albeit briefly, with some considerations which might seem unusual in a study of this kind though I believe that there is justification. One is Sitoni's possible Scottish descent which may have had something to do with the high position he seems to have enjoyed at the court in Madrid. Another is the state of society and technology in Lombardy and in the kingdoms of Castile and Aragon.

The general reason for all this is that there are still many obscure points about his personality and his work. So I have tried to leave as many clues as I can for future researchers.

The Man and His Work

Giovanni Sitoni was born in Milan on 11 November 1532. The Emperor Ferdinand I had named his father colonel of a cavalry regiment and therefore his family certainly had a good social and financial position. Philip II granted him citizenship of Milan. The books and documents I have been able to consult do not give the date of this grant nor his original nationality.[1]

Members of his family submitted a massive petition for a patent of nobility a few generations later in which they refer all the time to their Scottish ancestors. They assume that they were members of the Seton (or, in the Scottish form, Seytoun) family, one of the most important and notable in that kingdom, not only for its ancient origin, proven since 1150, and for its extensive properties, but also because many of its members won distinction in politics and the arts. Moreover, they bravely defended their country and their religious and political convictions. Several died on the field of battle and others were executed.

The aforesaid petition names Francesco Sitoni as the eldest of what we might call the Italian branch of the family. He was living in around 1485. As to the family in Scotland, the document claims that they possessed a castle near Edinburgh, on the River Esk and this appears to be true. Regarding the oldest ancestors in their kingdom, the petition names Alessandro Seton who, in 1330, being 'Captain General of the Navy, and in defence of his King, died after having combated gloriously', and also 'Giovanni Sitoni whom Robert II King of Scotland appointed as one of his six principal generals'. With regard to the reason for transferring to Lombardy, the petition states that 'This family moved for military reasons into France and Italy, following the example of other noble Scottish families ...'.[2]

The Alessandro Seton of the petition is almost certainly Sir Alexander Seton who is known of between 1311 and 1340. He signed the letter to the Pope in 1320 declaring Scottish independence and fought against the English. As for Robert II (1316–90), he was on the throne from 1371 to 1390 and was the first of the Stewart dynasty. During his turbulent and warlike reign, the only possible Seton appears to be Sir John Seton of Seton, father of Sir John Seton (d. 1424) and grandfather of George Seton, first Lord Seton (d. 1478).[3]

What interests us now are the two members of this outstanding family who had a connection with Spain (and we leave aside the exaggerations inherent in the majority of texts dealing with genealogies of those days).

George Seton (Fig. 1), fifth Lord Seton (1530?–85) and a very important figure at the court of Queen Mary, had to go into exile, possibly for being a Catholic. He returned to his country about 1561 and then in what might be a second exile—or perhaps as an extra-official Ambassador of the Queen—he went to Flanders probably about 1569. There he was in close contact with the Duke of Alba, who was governing the territory in the name of Spain. He tried to induce two Scottish regiments in the Flemish-dominated region to go over to the service of Spain in the hope of them later serving his Queen. He was discovered but was saved thanks to his fellow-countrymen who were officers in these regiments. It is said that the Duke of Alba helped him with money for this plot. But he is also supposed, whether before or afterwards is not said, to have had to earn a living driving a cart with four horses.[4]

His third son, Sir John Seton, Lord Barns, went to live in Spain when he was very young. Scottish texts say that Philip II made him a Knight of the Order of Santiago, gentleman of his chamber and Cavallier de la Boca, and granted him a pension of two thousand crowns to be transmitted to his heirs. These were very important honours, especially the first one, but I have not been able to find him in the register of the Knights of Santiago. As to the pension, that appears highly improbable, given the poor state of the Spanish Exchequer at that time. It is useful to establish at what period this took place. Sir John died in 1594 but, as with many other members of the family, the date of his birth does not seem to be on record. All that is said is that he died 'in the strength of his age, a young man', and James VI had made him return to his country before 1581. Thus he was probably born in about 1555 and was in Spain, more or less, between 1570 and 1580.[5]

All the above is of course, at this stage, tentative. Some Scottish historian may yet dismiss some if not all of it. And in any case, it would be pedantic to pursue exhaustively a matter which is, to a certain extent, of secondary importance.

Figure 1. George Seton, fifth Lord Seton. (By kind permission of National Portrait Gallery of Scotland.)

I am sure there is a foundation of truth in what is related about these persons, even though it has been embellished by the vanity of family members who wrote about them or had them written about. It stands to reason that if George Seton, a Catholic and a member of the highest Scottish nobility, had served Spain in the Flanders wars, Philip II would have known of it. And the Spanish King would have been glad to welcome a son of his.

As for Sir John, even if we make little of, or indeed set aside, the favours he may have received, I do not see any case for denying that he was at the Spanish court and thought highly of as a distinguished gentleman, especially considering not only his extraction but his personal merit, since on his return to Scotland the King entrusted him with some important tasks.

Returning now to Giovanni Francesco Sitoni, nothing is known about his technical training. In his time, most engineers did not study at a university although Giovanni may have done so, being of a family of high rank. Moreover, the literary style of his treatise is much better than many more or less contemporary ones.

His name appears in documents for the first time in 1556. At that time he was engaged in work on the Duomo at Milan. He is styled engineer and *agrimensor* (land-surveyor).[6] This last term, or some similar one, refers to the process of levelling; today we would say that he was fundamentally a topographer. This did not imply any demerit in relation to his being an engineer for geometry was particularly highly esteemed. This perhaps began with the influence of Pythagoras and of even more ancient hermetic texts. At the end of the same year he was named engineer extraordinary to the *Regia Ducal Camera* of the city of Milan.[7]

The College of engineers, architects and land-surveyors of the City and Duchy of Milan in 1563 created a commission to examine those who wished to belong to it and to decide on whether they were worthy of being admitted. There is no need to emphasize the importance which belonging to these corporations had given the rigid class structure of those times. Protests arose about the election of those who were to form part of the commission and appeals were even made to the King and the Senate. However, at last, six names were agreed upon and one of those chosen, in his capacity of *agrimensor*, was Sitoni.[8]

We come next to his activities as an hydraulic engineer in Spain which began in 1566. Indeed, he signed a contract that year with Gabriel Casato, President of the Milan Senate, by which he was to '... deal with the resumption of constructing work for the Imperial Canal (*Acequia Imperial*) of Aragon ...'. This irrigation canal was a very ambitious undertaking. It began in Navarre (Fig. 2) and was to reach Zaragoza, the capital of the kingdom of Aragon, with a length of around 80 kilometres. The people of Aragon had for a long time hoped for its construction which Charles V had commenced in 1530; indeed, he had been very interested in the work and had even overestimated the results that could be attained from it. In spite of this the ruinous condition of the Spanish economy—for many reasons, some of which I shall refer to later—had interrupted the work for a long time before Philip II asked for a report. Actually the work was only undertaken again two centuries later in the brilliant though brief period of the Enlightenment when the Canal was planned to reach the Mediterranean and be made navigable.

What is immediately surprising about Sitoni's contract for his first undertaking in Spain is the high salary he was offered; eighty gold escudos

Figure 2. The headworks of the Imperial Canal of Aragon as they appear in a nineteenth-century engraving. The building is of the time of Charles V.

paid in gold from the time of his leaving Milan until his return there, three months' salary for his travelling expenses, a supplementary costs allowance of a hundred escudos and an advance of four hundred.[9] Later we shall see that, in comparison with what engineers generally earned, this was exceptional. Especially is this so as there is no record of his ever having worked as an hydraulic engineer, although it must be supposed that he had done so and, since he lived in Milan, it was most probably on the network of navigable canals in Lombardy. These are very important in the history of technology. Only in the seventeenth century did the French build anything comparable. The first Lombardy Canal, which afterwards acquired the name of Naviglio Grande, reached Milan in 1257 and many others followed it. The ruling class there at that time was very interested in progress and, at the same time, Lombardy's geographical conditions were very favourable; it is on a plain which gradually slopes down towards the city. The soil is easily excavated and the seasonal variation in the volume of water in the rivers is moderate so that the spillways on the weirs were sufficient to control floods.

Two people in particular favoured this splendid hydraulic system, Francesco Sforza in the fifteenth century and Francis I of France in the sixteenth. Later on the lack of economic resources which I have to refer to so often very much decreased the work of canalization under Spanish rule. Nevertheless, some work continued even though it might only be for maintenance purposes, and thus there was employment for an engineer such as Sitoni. All the same, I must state that those who have very kindly helped me have not found any trace of him in the well-preserved Lombardy Archives. However, an exhaustive study of them has not yet been made.

Clearly work of this kind created a favourable atmosphere for the scientific development of technique. Engels's observation that 'all hydrostatics are a result of the need to regulate the flowing torrents in Italy in the sixteenth and seventeenth centuries' is true in general terms.

Sitoni's first sojourn in Spain should have been a short one since he delivered his report about the canal in August 1566 with a description which reached the conclusion that work ought not to be resumed given the lack of funds the country was experiencing. Yet he remained in Aragon until the end of 1568 without there being any sign of his undertaking any enterprise. The proof is that the King's Treasury paid him all that time. Then he returned to Milan.[10]

In 1569 he was in Spain again. In February, we have the first mention of the new work entrusted to him. Philip II approved an outlay of 2,000 ducats as a provision for material intended for the Colmenar Canal while 'waiting for the arrival of Francisco Sitón'. So we must now consider this hydraulic work, only of relative importance in itself, yet notable because of the engineers who were involved with it.

The kings of Spain for many centuries used to spend a considerable part of the year in places some distance from the capital, places which continue to bear the name of *Reales Sitios* (Royal Parks). There were several reasons for this, the chief one probably being their fondness for hunting.

One of these *Reales Sitios* is Aranjuez, a town about 60 kilometres to the south of Madrid on the River Tagus. Philip II put the two famous architects of El Escorial, namely Toledo and Herrera, in charge of the enlargement of an old palace (which has now completely disappeared, being replaced by another one constructed in the time of the eighteenth-century Bourbon kings) and he laid out some very beautiful gardens. He also planned and constructed a small reservoir which, in spite of its size, is called the 'Sea of Ontígola'. It was one of the first reservoirs to be once again used for recreation, in this case of the court, a custom which had been lost since the days of the Roman Empire.[11]

The Tagus had to be diverted from its course to water the Royal Gardens and this requirement coincided with the wish of the inhabitants of Colmenar de Oreja, some 23 kilometres to the north-west and relatively near the river, to irrigate their lands. Philip II took a special interest in this irrigation channel. In the same document, he mentions that the levelling of the future canal was entrusted to three engineers, though I believe that Sitoni was left to take the final decisions because Philip says '... and when Sitoni comes what is needed will be known'.

I will refer to one of the three engineers because his life is quite extraordinary even in those days when specialization was rare. His name was Ambrosio Mariano Azaro, a Neapolitan and therefore also a subject of the King of Spain. As a military engineer he took part in the Battle of

Saint Quentin when the Spanish defeated the French. Afterwards the King commissioned him to study the possibility of making the Guadalquivir navigable from Seville to Córdoba, that is to say for a stretch of some 160 kilometres along a large river. He was also ordered to do irrigation work on the Jarama River and on the Tagus as already mentioned. For this, he proposed a much more ambitious plan than the other engineers consulted.

But along with his engineering undertakings, he led a very complicated and extraordinary life. He was a secular theologian at the Council of Trent, an important figure at the Court of Poland and a Knight of Malta. He spent two years in prison accused of a murder, of which eventually he was acquitted. He became a hermit in Andalusia. Finally, he entered the Order of the Discalced Carmelites under the influence of his friendship with Teresa of Ávila. Many stories are told of him, some of which are very odd indeed.[12]

The King was rather impatient for Sitoni to arrive and take charge of everything. This is shown by his following the stages of Sitoni's journey carefully. He writes, 'he will come in the galleys which are expected any day'. And a little later, '... there is news of his arrival at Zaragoza, he has been ordered to come, then everything will be entrusted to him, and he will make the most fitting decision'.[13] The King seems to have had great confidence in Sitoni, which is strange in so far as Philip II had many engineers in his service, some of them famous ones, and the canal was not very important.

Those who made reports before Sitoni arrived were all in agreement, having seen the weirs of several mills, that the best one to be used as intake for the canal was that of Buena Mesón. In fact, this weir still exists today and is used for irrigation, the present structure probably superimposed over the old one.

Sitoni arrived, studied the work and issued his report. It is in good Spanish, like that of Ambrosio the Hermit, and not surprisingly since the Spanish language was very highly esteemed among cultured subjects of the Spanish Crown, especially if they had studied at a university. It is curious that his prose is better than that of his Spanish colleagues and that of Juanelo Turriano, who also took part, as will be seen later. But these colleagues of Sitoni's certainly came from a lower social class.

Sitoni's report[14] is technically accurate. It allows for an economic comparison of the different proposals very much as would be done today. It considers the intake at five of the weirs available: in each case it states the length of the channel and the area to be irrigated, distinguishing the part which was the King's property from that of the inhabitants of the villages and, finally, it calculates the cost. I have divided costs by areas and it turns out that Sitoni chose the second alternative, taking economy only into account: but, apart from that, it apparently had technical advantages.

The work began in 1570 under Sitoni's direction, the intake being located in the Sotomayor weir, which was not the one proposed in the first place. But in 1571 difficulties arose. The governor of Aranjuez ordered the work to be examined by Juanelo Turriano and two Spaniards because it seemed that the channel was running the wrong way and a plan to make a new one was needed.[15]

Turriano was the most famous of the engineers who worked in Spain in that century. He also won distinction, among other things, as an astronomer and as the designer of two planetary clocks which were considered unsurpassed. The first of the famous waterwork systems (*ingenios* or *artificios*) which he planned and built to elevate the water of the River Tagus to the Alcázar in Toledo had started running two years earlier.[16]

The results of the inspection made by Turriano and Jerónimo Gil and of a separate one by Benito de Morales were sent to Juan de Herrera and are relatively unfavourable. They proposed changing the intake and the spillway which they considered inadequate and the cause of water passing over the dam when there were floods. With their report, Turriano and Gil submitted a plan which unfortunately I have not been able to find.

Whether or not any notice was taken of these criticisms, Sitoni went on directing the work until, in December 1572, the King gave him leave to go '... into our state of Milan, of which you are a native, for certain business you need to attend to ...'.[17]

This same year, 1572, a significant incident in Sitoni's life took place. Martin Gaztelu, secretary to Philip II and one of his most efficient aides, wrote, by the King's command, to Juan de Herrera asking him to '... let me know your opinion of the ability of this Sitón in the matter of channels since, though there are others here of his profession ... His Majesty does not know if they have the same practical experience as the aforesaid Sitón in this matter of irrigation nor whether he has quite finished the channel he was making in Aranjuez ...'. The implication of this is that Sitoni was considered a good engineer and, on the face of it, better than the others.

Herrera returned the letter to Gaztelu, jotting on the back a very harsh judgement which I shall quote here, omitting repetitions: 'When he came from Italy His Majesty asked me what he knew and I replied that he knew nothing which could be useful to his Majesty and that he did not know about the work His Majesty sent for him for, that is to say, levelling. This could be seen by the instrument he brought with him which was inaccurate and he knew nothing of construction work ... I believe the Count of Chinchón found this out by experience. I think all this was thought suspicious at the time, but I told the truth.'[18]

This can be explained in many different ways. One is that Herrera, who not only knew about topography but had invented instruments—including one intended to solve the famous problem of determining geographical longitude—felt humiliated at not having been consulted. Or it could of course have been a reaction against foreigners, though this is improbable, as xenophobia was not greatly developed at that time and, besides, Lombardy belonged to the Spanish crown, just like Castile.

I suggest another explanation. For this we must remember that the financial position of engineers and architects was relatively good with the former slightly better off than the latter because there were fewer of them. That difference can be understood because more money was spent on buildings than on public works, even including fortifications. This may seem surprising but it happens as capitalist systems begin to develop. Very many of the buildings being constructed in Spain at that time were religious ones, that is to say churches and convents. Indeed, the Church received a disproportionate percentage of the national income, part of which it spent aiding the very poor it had helped in the first place to create, while another part was invested in those works of architecture and art which have contributed to this period being called the Spanish Golden Age.

On the other hand, the crown, ruined by an irrational foreign policy, together with inflation and an ever-growing interest on debts, had not many funds at its disposal for public works. What has been said above about the Imperial Canal is a good example. Owing to the late development of capitalism, the need for funds could not be met by private money even when something was attempted, though not much, especially in the Mediterranean region.

The nobility, another rank of society greatly favoured in the division of wealth, was also a good client for the architects, though not perhaps to the same extent as in Italy and France. The bourgeoisie only began very late to take any part in the life of the nation.

I am now going to give the annual salaries earned by Juan de Herrera. It must be remembered here that he is considered to be one of the great figures of world architecture and that he was a good friend of King Philip II. His importance as an architect was so great that his new, personal style—'herreriano' as it is called in Spanish—can be found in many places in Spain in buildings not designed by him. As assistant to Juan Bautista de Toledo, who began the work on El Escorial in 1563, he was paid one hundred ducats,[19] while Toledo never earned more than 420. When Toledo died in 1567, Herrera's salary was raised to 250 ducats per year. In 1569 he obtained a position at court as well, which was not connected with his profession, and between the two he made 650 ducats. In 1577 he reached the height of his fame and influence, and at the same time achieved his maximum salary as an architect: 800 ducats. Though he never ceased working altogether, the *juro* (annuity) of 1,000 ducats, which he received from 1587 onwards, was in recognition of his long and glorious career in the service of the crown.

Juanelo Turriano never earned more than 400 ducats per year, though he had an income of 200 ducats from some property in Milan, a grant which had been given to him by Charles V as a reward for having planned and constructed one of the planetary clocks referred to above. But that was an extraordinary payment that cannot be reckoned in a comparison with what others earned.

We find that Herrera in 1572, when he was consulted about Sitoni, was the designer of one of the most famous monuments in Europe and of many other structures, some of them works of engineering. He even found time to devote to philosophy, science and mechanical inventions. He was then earning 650 ducats a year. Meanwhile, Sitoni was making 960 when he worked ('was engaged in matters of his profession' as the Royal Order says), and received 600 ducats when he was doing nothing. Nevertheless, he expressed dissatisfaction on being informed about this reduction. And, if we go back to the time when he arrived in Spain, in 1566, he was earning the first figure mentioned, 960 ducats, while at that same time Herrera, whose opinion was that Sitoni knew nothing, received 100 ducats.

One must be careful not to modernize historical data. All the same, the enormous difference of salary is incomprehensible, especially as Sitoni appears to have only made a report of a general nature and later levelled, planned and directed—expertly or not, as the case may be—a not very long irrigation channel. I do not think the explanation can be that Sitoni was also engaged on other work of which no record remains because the age of Philip II is proverbial for its bureaucratic control and for the files which were kept of every little thing.

I cannot find any other explanation than the one already put forward, namely that he was ranked as a nobleman because of his descent from the Scottish Setons and, for that reason alone, was well thought of at court and highly paid accordingly. Unless, of course, there was some reason of a personal nature which we shall probably never know. Historical anecdotes are usually only about important figures.

This apparently very comfortable life continued until the end of 1579. At that time the King named him engineer of the *Regia Ducal Camera* in Milan. For the services he had rendered, he had the right to choose which of his sons should succeed him in the post. Altogether he was some eleven years in Spain and, apart from the report on the Imperial Canal, there is evidence of only two works by him during the whole of this time.

The first one was the Colmenar Canal. It is not known for how long he was there but by 1575 the job was being directed by Jerónimo Gil, whose name has appeared above. Gil was a distinguished architect and sculptor in wood. He had previously worked on the Toledo Alcázar and on the Palace and Chapel of Aranjuez.

Sitoni's second work is even more mysterious. The only Spanish reference to it says that 'the King, having issued a commission to his Counsillor Martín(?) Juan Franquesa to draw out an irrigation channel using the water of the Segre River in Catalonia and a dam between the towns of Pons and Oliana, commanded Sitoni to make a survey of the land and the necessary plans, at the orders of the aforesaid Franquesa. But none of this was put into effect.'[20] Certainly in the neighbourhood of the Catalan city of Lérida, irrigation channels, some of very ancient origin, water the land along the Segre and Noguera Ribagorzana rivers. But though historical references to these irrigation channels exist, I have not found any mention of Sitoni in them. The most probable explanation is that Sitoni only made a report, preceded by levelling, but that nothing was constructed because of the lack of funds, as with the Imperial Canal.

In the treatise Sitoni wrote, which I shall refer to later, things are presented in this way: '... by order of His Majesty, in the year 1578, between the town of Noriola and that of Coscó, in the Kingdom of Catalonia, I planned using the waters of the Segre, a notable and very important river in those parts, to supply the countryside called Urgel between the city of Lérida and the town of Tárraga in the same Kingdom ... which covers a plain of more than 150 square miles.' The area referred to seems very small and perhaps this is a mistake in the text.

When he was in Italy, human nature being weak, he must have magnified the importance of the part he played in Spain. A dictionary of illustrious Milanese says: '... he served the Catholic King with daring inventions to such a point that, to the astonishment of Nature and Art, canals were cut through the peaks of mountains to water whole provinces, which thus became more productive and populated'.[21] Unfortunately for Spain this never took place.

When he returned for good to his country, he was sixty-seven years old. During his closing years in Milan he continued to be active, but my impression is that he did not undertake anything very significant. However, he did have an important social position as a high-ranking servant of the Duchy and probably because he had made his fortune. I will give a summary of what I have found out about him.

In 1603 he formed part of a commission to consider whether the pillars of one of the façades of the Duomo should or should not have pedestals. It was decided that they ought to have them, but the matter was still being discussed in 1607,[22] though I do not know if he was any longer concerned with it. Curiosity led me to go to look for these pedestals and they exist on the eastern façade.

He also took part in the building of the fortress of Fuentes (Fig. 3), a very important Spanish stronghold on Lake Como. It was given the name of the Governor of the Milanese from 1600, Pedro Enriquez de Acevedo, Count of Fuentes, who ordered its construction. He was a remarkable man whose merits have been recognized even by historians very critical both of Spanish domination and of governors who were military men.[23] Work began in 1603 and the fortress rose to an impressive size. But it is not clear what Sitoni did there either. The dictionary quoted from above states, just after the tall stories about his Spanish canals: 'And (*the works*) he performed in the milanese were no less remarkable; in fact, he was the Count's advisor ... in the building of that very well fortified citadel ...' But only one document is known in which he is referred to by the name of Francesco Sitoni, though certainly the same person is meant, together with several others who measured the fortress in 1608.[24] Thus once again he was involved with topographical work. Another passing reference to him confirms this: he evaluated the necessary costs of work on the Monastery of S. Vitore[25] (which it is certain required a previous cubature).

Figure 3.   The fortress of Fuentes. From an old engraving.

On the other hand, certain references, the last ones I will give, seem to agree with my idea that in these years he was a member of the patriciate of his native city. Mention is made of his collecting several antiques and reproductions of the most beautiful statues of that day, though it is not explained if he did so for himself or for somebody else.[26] Another reference is to his being asked, along with others, for his opinion about adapting, for the official use of the Duchy, the Palace which had belonged to Tomaso Marino, Duke of Terranova, and which the public exchequer had taken over.[27]

On 11 August 1608 he died at seventy-five years of age.

The facts I have been able to assemble, even with the advantage of much unselfish help which I shall refer to later, only allow us to form a more or less approximate idea of what Sitoni was like. But I will try to set it out clearly.

The posthumous praises in Italian books of his great engineering feats I consider to be untrue. They come from the patriotism which, in a sense, has its origins in Italy during the Renaissance, and which always grow when commenting on periods of foreign domination.

He belonged to the ruling class, being in the service of the crown both in Spain and 'n Lombardy. Besides, in Lombardy he was an outstanding member of a fairly powerful corporative organization. In Spain, as I suppose, being a courtier helped with his promotion and it is certainly probable that being considered a nobleman did so too. Of course I may

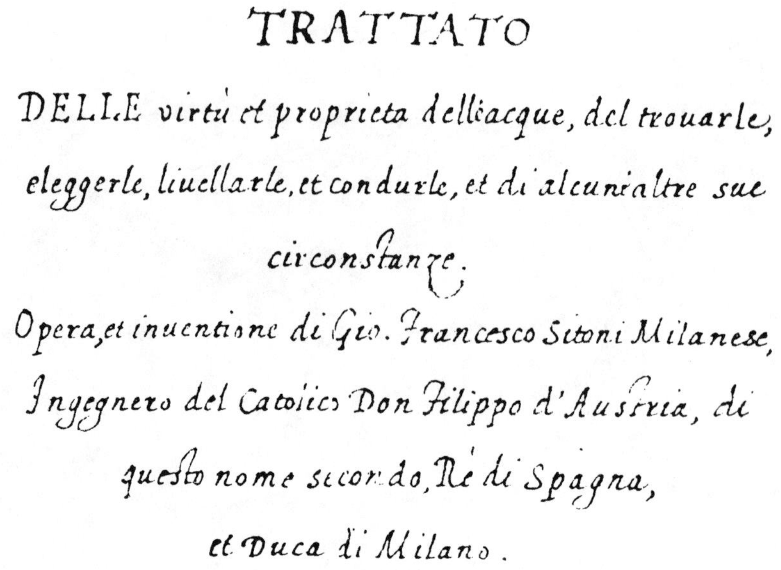

TRATTATO

DELLE virtù et proprieta dell'acque, del trouarle,

eleggerle, liuellarle, et condurle, et di alcun'altre sue

circonstanze.

Opera, et inuentione di Gio. Francesco Sitoni Milanese,

Ingegnero del Catolico Don Filippo d'Austria, di

questo nome secondo, Rè di Spagna,

et Duca di Milano.

Figure 4. The title page of the Sitoni codex.

be mistaken, and I recognize that in the Italian accounts, for all that they are so enthusiastic, no great importance is attached to his nobility, though it is admitted.

On the other hand, Sitoni was well thought of as an engineer, even though the objective facts we know about his work show that it was modest. Apart from his directing the work in Colmenar, I can only quote his reports, topographical plans and some estimates.

But Sitoni's importance is considerably increased because, in those years in which he was neither idle nor too hard at work, he was writing, perhaps in Spain and certainly in Milan. I shall now deal with this.

The Codex

In 1935, the Burndy Library was founded in the American city of Norwalk, Conn. It is devoted to the history of science and technology, and very important books and manuscripts are kept there. I have, for many years, enjoyed the honour of being a good friend of its founder who is still its President, Bern Dibner.

I found there a long time ago a codex (Fig. 4) in Italian entitled 'TREATISE/OF THE virtues and properties of waters and how to find them/move them, level them and channel them, and some other of their characteristics/Work and invention of Gio. Francesco Sitoni Milanese/Engineer of the Catholic Don Philip of Austria, second of/this name, King of Spain,/and Duke of Milan'.

Before commenting on this codex, it must be said that the appearance in the sixteenth century of the first technical monographical treatises was an important contribution to material and intellectual progress. Two of them were printed and are justly famous.

In 1540, *Pyrotechnia* appeared in Italian shortly after the death of its author Vanoccio Biringuccio of Siena. It deals with metallurgy. In 1556 Georg Bauer, a Saxon, published *De Re Metallica* in Latin; he also latinized his own name so that posterity remembers him as Georgius Agricola; his field was not only metallurgy but also mineralogy and, especially, mining.

But at that time no treatise on hydraulic works was printed. This omission is now being corrected, and the author of this paper has had the good fortune of being able to contribute to this work.

The men who wrote the three works I am now going to refer to acquired their experience in Italy and Spain, which is not to be wondered at. I have already referred to Lombardy, and in other states there were remarkable works and machines. In Spain too, hydraulics developed greatly and in a form we could call complementary. As there was less rain in a great part of the territory, storage dams were made, a type of structure largely forgotten for centuries in all countries. Systems of engineering (civil and mechanical) were altered in order to use the flows from the rivers as much as possible.

The most ancient treatise on hydro-technology of which there is any record is a manuscript that has been lost. Indeed, Girolamo Cardano in his book *De rerum varietate libri XVII* (Avinione, 1558),[28] p. 69, after explaining how a windmill works, writes: 'Besides this, anyone interested in a detailed description and a drawing of the machine, should go through the book by the Spaniard Jerónimo Giraba (*Hieronymi Giraua Hispani*) which gives a very complete account of the aforesaid subject.'

Until I came across this passage from Cardano, I believe the work had only been mentioned in modern times as having been in Juan de Herrera's library. The inventory made when he died gives the Spanish title of the book, which in English is: 'Declaration of the use and fabrication of water instruments, mills and other things, in manuscript, by Gerónimo Xiraba.'[29] That was not enough to form a clear idea of the work since it may have been a single copy, falsely attributed because there is no record of the author, a scientist of certain importance, being interested in technology.

Giraba was a cosmographer in the service of Charles V and he lived a substantial part of his life in Flanders and in Italy. In collaboration he translated, in 1553, a book on geometry by a French writer, Oronce Fine: the manuscript, with beautiful drawings, still exists. And he published in Milan (1556), a work called *Dos libros de cosmographia* (Two books on Cosmography), which is really a compendium of geography. To be sure, his life and works have not yet been studied as they should be.

In my opinion, Cardano's statement only makes sense if the treatise could be consulted, perhaps even obtained, with no great difficulty. Possibly a copy or so may be found in the libraries and archives of Italy, as it apparently came to be known in that country and was perhaps even written there. There is nothing strange in Herrera possessing it since he usually had books sent to him from abroad. As to its being quoted in a passage referring to windmills, this does not infer any mistake. Windmills were one of the two alternatives for obtaining energy, and the hydraulic mill predominated in both countries and even more in Spain, the author's country.

Not only the title but Cardano's recommendation of the book seem to show it was important, at least on the mechanical side of hydro-technology, while the mention of 'other things' provides support for this supposition. The *post quem* date does not seem to me to be very long before the one stated, 1558, so we are not far removed from the period when Giraba was most active.

The existence of the second treatise has been known for centuries but, until recently, although it was frequently referred to—though sometimes with serious errors—it was rarely read. It is a long compendium of the nature of every technique connected with water and is kept in the National Library in Madrid. It had always been attributed (falsely, as I believe I have shown) to Juanelo Turriano, the Lombard engineer to whom I have already referred for other reasons. For many years I have studied this treatise and worked to have it published, something which was finally realized in 1983. I gave the author's name as 'Pseudo-Juanelo Turriano' and the work appeared under its title 'The Twenty-one Books of Devices and of Machines'. The complete Spanish text was given and most of the edition includes an introduction in Spanish; however, in 500 copies my introduction is in English.[30]

The pseudo-Turriano is of great importance and I have equated it with the works of Biringuccio and Agricola which have been quoted. The text has about 350,000 words and there are 440 drawings of high technical quality and, in many cases, artistic quality as well.

My introduction is divided into three parts: the history of the codex, a description of it and the problem of its attribution. And the fact is that the name of the author is still not known, though I and others are trying hard to determine it; I shall refer to this again.

'The Twenty-one Books . . .' display a certain similarity with some modern works on the same subject as the following list will confirm:

1.   HYDROLOGY
1.1.   Hydrological cycle. Properties.
1.2.   Prospecting.
1.3.   Quality.
2.   TOPOGRAFICAL INSTRUMENTS (of necessary use for what follows)
3.   HYDRAULIC CONDUITS.
3.1.   Deposits.
3.2.   Open channels: aqueducts.
3.3.   Subterranean conduits.
3.4.   Crossings at different levels. Siphons. .
3.5.   Transport from springs. Water supplies. Auxiliary machinery.

4.   WEIRS.
5.   CISTERNS.
6.   BRIDGES.
6.1.   Of boats.
6.2.   Of wood.
7.   MATERIALS.
7.1.   Wood, stone, brick, binders.
7.2.   Trees.
7.3.   Stones: extraction and cutting.
7.4.   Stones: quality. Bricks and tiles.
7.5.   The materials for making concrete.
7.6.   Bitumens.
8.   ENERGY AND ITS USE.
8.1.   Mills.
8.2.   Systems of elevating water.
8.3.   Uses of energy (industrial and agricultural).
9.   STONE BRIDGES AND THEIR FOUNDATIONS.
10.   PORTS.
10.1. Maritime works.
10.2. Fortification and equipment.
11.   VARIOUS.

Of course, observations should be made about this supposedly proper order. It is correct until section 5. And then after dealing with bridges (6), it covers stone ones (9) separately because the notion of how they should be built is different and they need a different type of foundation; these last are to a great extent like the ones in maritime work, which are referred to immediately afterwards. It reverts to a relatively normal order with materials followed by mills. Afterwards there comes the use of energy and ports.

It should be noticed that a number of techniques, which would be considered separately today, are referred to because the author does not exclude anything related to water, not even ports. Topographical instruments and materials are dealt with because they need to be used by a hydraulic engineer; bridges are included because they cross rivers.

And finally, I estimate the date of the writing and the drawings as probably between 1590 and 1600: conceivably it could have been written slightly earlier. This dating is necessarily approximate and subject to revision.

The Sitoni codex is much shorter: 221 pages, comprising about 42,000 words. Figures are mentioned but they are missing and moreover there could only have been a few of them.

It begins with an index to which I shall refer again. Then comes the dedication to Simone Bosso of the King's Secret Council in the State of Milan, according to which he intended to communicate his experience to one or several of his small sons, in case they should later be drawn towards the profession of engineer or land-surveyor. He mentions the scarcity of treatises like this one, '. . . as Vitruvius and all the others who have written about this and whom I know of, have only written about inventions and the manner of choosing waters. But as to conducting them, making use of them and levelling them, they have hardly mentioned these . . .'. This suggests that Sitoni knew no similar work, but this is not of course certain.

At the end of the dedication the place and the year are given: . . . in Milan on the—of——1599. The codex is the work of at least two scribes, not very good ones, but legible. There are very considerable corrections to the text; deletions, additions and attempts to improve the style. It seems certain that they were made by Sitoni himself in preparing the final copy for printing. This hypothesis is supported by the division into chapters. There are 53 in the index—each one therefore being very short—and besides correcting his titles he removes several of them without adjusting the numbering. On the other hand, he does make the adjustment in the body of the work, though in an imperfect way. Finally, it would appear that the urge or the opportunity to publish the treatise was lost as the incomplete date confirms. Nor could there have been any other copies in this condition, although it is true that the title is quoted in a few books.

Returning to the problem of the attribution of 'The Twenty-one Books . . .', as I pointed out in my introduction to it, when it was published an important clue to the problem of authorship is that the Spanish includes many words of the dialect spoken in the kingdom of Aragon, where almost all the villages and towns which are referred to in the book can be found. The Italian places referred to are mostly in Lombardy, the others being in the Pontifical States. The techniques suitable to the two countries are also set down side by side. Many examples could be given, but I shall quote only that of hydraulic mills with a horizontal wheel, more suitable for dry countries, and also the systems used for transporting a boat from one river to another, only possible when the rivers have an abundant flow for a great part of the year.

These and other reasons led me to the conclusion that an unknown Aragonese author wrote the part of the treatise which is principally applicable to Italy, making use of a friend from Italy, who might probably be Sitoni. When I wrote this, of course, I considered that a different solution might be possible. Re-reading the Italian codex with greater care has increased my doubts. The very great importance of the manuscript which was attributed to Turriano made me think, on finishing my introduction (in August 1982), that it was very unlikely that anyone who had been involved with writing such a work should have left no trace of his existence, though not taking into account that even now great men die unknown. I am confident that quite soon we shall know who our author was and of course we must keep in mind the hypothesis that Sitoni, with others, may have taken some part in the preparation of the 'Twenty-one Books . . .'.

This paper will be completed with the indexes of both codexes which will help future researchers in a comparative study. Even if the conclusion is reached that there is nothing which allows a connection to be established between them, it is still very interesting to see how the second oldest known text on hydro-technology developed. Perhaps one day it will be published.

For the division into volumes and books and pagination of the first manuscript, I refer to my General Introduction.[31] The numbers for Sitoni's treatise are those of chapters. The titles in the first are accurate. Those in the second may have transcription errors because the corrections noted above sometimes make the manuscript difficult to read.

Pseudo-Juanelo Turriano

Qualities of the waters and of their properties and of their generation (or source).
Of the effects of the waters and of what they do within the earth.
Of the fattiness of the waters.
Of the signs for finding water which is hidden within the earth.
Of the signs for finding water and which are the true ones.
Of the experiments to be made to find water.
How we can know (whether) water is good or not.
Of the levels and their forms (for these constructions).
Of several ways of carrying water and (of the plans to be made to build) aqueducts.
Galleries as they should be made and how to make irrigation channels to carry water in several ways.
To carry waters which pass some beneath others.
Of the differences there are in carrying water from springs.
Of the different kinds of weirs.
Of cisterns (cisternas y aljibes). How they are made in different ways.
Of boats that are used instead of bridge(s). And of (other) bridges.
Of bridges of wood only.
Of wood and stone and (when) they are cut and how the stones are quarried and how to make concrete and gypsum and bricks in several ways.
Of trees in short.
Stones in general and at what time they should be cut from the quarry and in what state and at what times they should be worked upon and which are most easily broken and which are most lasting for the work.
Of the quality of stones and the manner of making bricks and tiles and other things of earthenware to adorn buildings.
What quality of stone is best to make concrete.
Of different kinds of mills (and horse mills).
Different means of sifting flour.
Fulling-mills and oil-mills, and of various kinds of artifices of the same quality to draw out water to make alums and saltpetre and to wash wools and cloths.

Of how to make the piers of stone bridges of various kinds.

Of sea buildings and how they should be made and set up in various ways.

Of making defences for ports so that navies cannot enter.

Of divisions of water and of islands, and other things (concerning) water.

Sitoni

1   Of the virtues and properties of waters and of how they are a very necessary element for life and generation among men.

2   Which are the good natural waters to drink and how this can be tested.

3   Of medicinal waters.

4   Of the indications and signs, by which waters can be found beneath the ground, how they can be tested, and how they can be chosen.

5   How and in what way a remedy may be found for those waters which have some unhealthy or bad quality.

6   How wells are dug and built, when the waters have been found and chosen.

7   How springs are dug so that, once their waters have been found and tested, they can be levelled before they are conducted.

8   Of the quality the waters should have for irrigation and how the coldness of the springs is tempered.

9 + 11 to 20
    Eleven chapters devoted to the manner of levelling and the instruments to be used.

21 Of the duty Masters and Engineers are under to consult with the Lords of the land about the works, constructions and their likings.

22 How water conductions are planned and why it is fitting that the outlets used for irrigation should not be deep, nor near hills.

23 How waters are drawn from streams to drink and irrigate.

24 How covered water conduits for irrigation are made and kept.

25 How weirs or dams in rivers are made to hold in the water which is to be conducted for irrigation.

26 How and where the main sluice-gates of headraces (mill-races?) are set, and how they are made so that they can be protected against the flooding of their rivers.

27 How spillways are made and which of them are more suitable for conduits, headraces (mill-races?) and, particularly, navigable waterways.

28 How bridges over the conduits are made and the difference between them.

29 How canals are made and ... what system is best and most lasting to make waters pass above and below other waters.

30 How it is known if the waters flow through all the parts of a conduit and if they maintain the same quantity and level.

31 How and at what time of the year running waters are measured so as to know their exact quantity.

36 What should be done and tested to know the difference of the quantity of water that passes through a limited orifice which has greater or lesser ...

37 How and in which way a nozzle shall be made ... so that a set quantity of water can be obtained from it in two or more days ... or in a whole week.

38 A rule for dividing the time for each watering.

39 How the drawing of a quantity of water can be controlled by hours and days of the week ...

32 How drained water can be measured.

33 The differences that exist between water measurements in the State of Milan.

34 How and with what, nozzles are made to draw out a certain quantity of water from other water conduits and at which time of the year it is put to use.

35 Of all the forms of the nozzles described in this treatise, all of them having the same area, the round ones always provide for a greater quantity of water.

41 How a certain quantity of water can be drawn from a conduit which is crossed to draw at different levels ...?

42 How and in how many ways the waters divided for irrigation are distributed ...

43 How land is levelled out for meadows or for irrigation and the advantage of doing it properly the first time.

44 Of the width conduits? meadows should have and how the channels for irrigation should always be set in the highest part of them.

45 How headraces (mill-races?), large and small, and the drainage of meadows are made and which kind of plants it is best to plant on banks to strengthen them.

46 How the land wanted to be made into a meadow is to be divided and what errors must be avoided.

47 Of the devices to enclose and keep waters in conduits and of the fabrication and use of them.

I would like to thank all who have helped me in the preparation of this paper and especially the following:

In England: Charles David Ley and Valerie Smith.

In Italy (Milan): Gabriella Anedi, Prof. Giulio Bora, Licia Carubelli, Marina Messina and Dr Giuseppe Scarazzini.

In USA: Dr Bern Dibner.

In Spain: Antonio de Juan, Prof. Sebastián Mariner and José L. de la Peña.

Notes

1. Archivio della Lombardia, Milan. Araldica, Cartella 120 *Fatto Genealogico con Documenti Comprovanti le qualita Generiche e Specifiche con il Blasone Gentilizio della Nobile Famiglia Sitona di Scozia.*

2. As note 1.

3. *The Dictionary of British National Biography*, Oxford UP, reprinted 1921-2, Vol. XVII, p. 1205, the entry for SETON, GEORGE, first Lord Seton (died 1478).

4. As note 3, pp. 1206-8. George Seton, Advocate, *A History of the Family of Seton during Eight Centuries*, 2 vols., T. & G. Constable, Edinburgh, 1896, Vol. 1, pp. 173-5.

5. As note 3, p. 120. As note 4, Vol. 2, pp. 621 ff.

6. *Annali della Fabbrica del Duomo di Milano*, Milan, 1881, Vol. 4^0 year 1556, pp. 22-3.

7. Archivio della Lombardia, fondo autografi, ingegnieri 1556: Cart 86, Fasc. 38.

8. P. Mezzanotte, *L'architettura da F.M. Richino al Ruggeri* in *Storia di Milano*, Milan, 1958, Vol. XI, p. 441.

9. Archivo General de Simancas (AGS). Casas y Sitios Reales (CSR): leg. 253, fol. 75 (8.2.1566).

10. E. Llaguno and J.A. Cea Bermúdez, *Noticia de los arquitectos y arquitectura de España*, Madrid, 1829, Vol. III, p. 19.

11. N.A.F. Smith, *The Heritage of Spanish Dams*, Madrid, 1970.

12. J.A. García-Diego, *Sobre aventuras de ingenieros italianos renacentistas*, Lecture in Real Academia de Ciencias, Madrid, 1984.

13. Archivo del Palacio Real (APR). Cédulas Reales (CR): Vol. 3, fol. 109 (18.3.1569). Id.: Vol. 3, fol. 118 v. (8.6.1569).

14. As note 9, AGS-CSR: leg. 252, fol. 112 (1569).

15. As note 13, APR-CR: Vol. 3, fol. 219-20 (20.1.1571).

16. J.A. García-Diego, *Los relojes y autómatas de Juanelo Turriano*, Madrid-Valencia, 1982.

17. As note 13, APR-CR: Vol. 3, fol. 419 v. (1.12.1572).

18. A. Ruiz de Arcaute, *Juan de Herrera Arquitecto de Felipe II*, Madrid, 1936, pp. 176-7.

19. A ducat was equivalent to one scudo.

20. As note 10, Vol. III, p. 20.

21. F. Argelati, *Bibliotheca scriptorum mediolanensium*, Milan, 1745, Vol. II, p. 1419.

22. C. Bocciarelli, *L'attivita di Francesco Maria Richino nel Duomo di Milano ...* Congress in Museo de la Ciencia y dela Técnica, 8-12.10.1968: Vol. 1^0, Milan, 1969.

23. Raccolta Ferrari, XVI. Quoted by M.L. Gatti, *Cose militari: Contributo alla studio della fortificazioni in Lombardia durante el periodo della dominazione spagnola*, 1970, p. 178. As note 20. P.A. Baldrati, *La fortificazione spagnola nell'alto Lario*, 1970, pp. 213-18.

24. Raccolta Ferrari, as note 23.

25. C. Baroni, *Documenti per la storia dell'Architettura a Milano nel Rinascimento o nel Barocco*, Roma, 1968, p. 214.

26. A. Scotti, *I disegni alessiani nelle collezione milanesi*. Atti del convegno 'Galeazzo Alessi e l'archittetura del Cinquecento europeo', Genova, 1974, p. 472. Genova, 1975.

27. Archivio della Lombardia. Fondo Uffici regi. Cartella 748. 1608.

28. There is a previous edition (Basilae, 1557) which I have not consulted.

29. J. Cervera, *Inventario de los bienes de Juan de Herrera*, Valencia, 1977.

30. Pseudo-Juanelo Turriano, *Los Veintiun Libros De Los Ingenios Y De Las Maquinas*, Prólogo, José A. García-Diego, 2 vols., Ediciones Turner, Madrid, 1983.

31. As note 30, Vol. 1, pp. 22-4.

Information on Engineering in the Works of Muslim Geographers

DONALD R. HILL*

Historians of engineering concerned with periods before the sixteenth century are acutely conscious of the scarcity of evidence. It is probably the case that there is more material available for medieval Islam than there is for the classical/Hellenistic period or for the early Middle Ages in Europe, which is not to imply that the documents for the Muslim world are abundant. It is the purpose of this article to indicate how our fund of information on Islamic engineering can be increased by studying the works of Muslim geographers and travellers. However, to begin with it will be useful to survey the material that is *specifically* concerned with engineering.

First, there are a number of treatises concerned with *hiyal*, an Arabic word usually translated as 'ingenious devices' or 'automata', although its meaning can be very much broader, to the extent that it can embrace all aspects of mechanical engineering.[1] The earliest surviving such treatise was written by three brothers known as the Banū (sons of) Mūsà about AD 850 in Baghdad.[2] Most of the devices described are trick vessels but there are also a number of fountains, self-feeding and self-trimming lamps, and even a clamshell grab. The Banū Mūsà's work, although derived from their Hellenistic predecessors, goes well beyond them, particularly in the masterly use of small variations in aerostatic and hydrostatic pressures and in the use of automatic control systems using conical valves. Late in the tenth century, Abu 'Abd Allah al-Khwārizmī compiled a scientific encyclopaedia called *The Keys of the Sciences* (*Mafātīḥ al-'Ulūm*) in which the components of *hiyal* are described, with instructions for their manufacture.[3] In Muslim Spain in the eleventh century, a certain al-Murādī wrote a treatise on water-clocks and automata machines. Unfortunately, the only surviving manuscript of this work is so badly defaced that it is impossible to determine precisely how any of the machines worked. It is, however, a document of the utmost importance since it includes descriptions and illustrations of multiple gear-trains incorporating segmental gears and epicycles. These are the first-known examples of complex gear-trains intended for the transmission of heavy torque.[4] *The Book of the Balance of*

* I wish to thank the Royal Society for their generosity in awarding me a grant to further my research into the history of technology.

Wisdom (Kitāb Mizān al-Hikma) was written by the great physicist al-Khāzinī in 1121. The book deals with a number of subjects including the specific gravities of liquids and solids, theory of gravity, measurement of gravities and applications of the balance. In the Eighth Treatise, al-Khāzinī describes the construction and use of steelyard clepsydras for the measurement of time. This is mainly of interest for the close attention he pays to accurate dimensioning and the close specification of materials.[5]

In a book called *On the Construction of Clocks and their Use*, written in 1203, Ridwān ibn al-Saʿātī describes the monumental water-clock built by his father by the Jayrun gate in Damascus. Despite some obscurities and inaccuracies due to Ridwān's unfamiliarity with engineering, the book tells us a good deal about methods of manufacture and construction.[6] It cannot be compared, however, with al-Jazarī's magnificent book on machines, written in Diyar Bakr in 1206. This contains descriptions of water-clocks, candle-clocks, automata, fountains, water-raising machines and a few miscellaneous devices. It is perhaps the most remarkable engineering document to have survived from any cultural area in pre-Renaissance times. It contains descriptions of techniques and detailed instructions for the construction of machines, from the smallest components to the completed assembly. Some of these techniques, components and machines are already known to us from the works of Hellenistic and earlier Arabic engineers, but a number appear for the first time in al-Jazarī's book and some of them may indeed be his own inventions. Perhaps the most important aspect of the book, however, is that the devices were described with the declared—and successful—intention of enabling later craftsmen to reconstruct them.[7]

Still on the subject of *hiyal*, there are a number of manuscripts extant only in Arabic which are compilations of Greek, Hellenistic, Byzantine and Islamic sources. These separate elements are not identified, and the treatises are usually ascribed to Greek or Hellenistic writers. For example, there are a number of manuscripts in Arabic attributed to Philo of Byzantium (fl. *c.* 230 BC), but all of these contain Islamic additions. Carra de Vaux' scholarly Arabic edition of Philo with French translation does not utilize all the 'Philonic' material, nor does it always succeed in making a convincing distinction between Philo's original work and the Islamic additions.[8] Another document, a treatise entitled *On the Construction of Water-clocks*, is attributed to Archimedes. I have published an English translation of this work, based upon the three known manuscripts, but my conclusions about the origins of the various sections are tentative.[9] There are also several Arabic manuscripts of a treatise on a musical automaton attributed to Apollonius 'the geometer, the carpenter'. On present evidence, this could have originated almost anywhere in the eastern Mediterranean area at any time from 250 BC to AD 750! This treatise was translated into German by Eilhard Wiedemann but the translation, admittedly based upon a very obscure text, is far from satisfactory.[10] The remaining manuscripts in this category have yet to be published, or even studied in detail.[11] While the value of this type of material is diminished by the uncertainty about origins, they are nevertheless of value in helping us to trace cultural diffusions. And, of course, in those cases where the origins can be established with some credibility, our stock of knowledge about engineering attainments in a given cultural area is thereby increased.

Two valuable treatises on hydraulic engineering have come down to us, both written in Iraq in the eleventh century. The first of these, by an Iranian called Muhammad b. al-Hasan al-Hāsib al-Karajī, is called *The Bringing to the Surface of Hidden Waters (Inbāt al-Miyāh al-Khafiya)*. It contains a section on surveying instruments and their use, methods for detecting underground sources of water and instructions for the excavation of *qanats*—the underground conduits still in use today in Iran and parts of the Arab world.[12] The other treatise is concerned with the construction and maintenance of canals in the Iraqi irrigation system: this also contains descriptions of surveying instruments and their use. The various types of water-raising machines are listed, and although very little is said about their construction, there is a very interesting passage in which the output of each machine is given, together with the area of crops each could irrigate in winter and in summer. The final section deals with quantity surveying, i.e. how to calculate the quantities of excavation for canals and hence derive the cost of the works in labour and in money.[13]

One category of writing that deserves more attention from historians of technology is *hisba*, a word that is not easy to translate. On the one hand it means the duty of every Muslim to 'promote good and forbid evil' and, on the other, the function of the person who is effectively entrusted in a town with the application of this rule in the supervision of moral behaviour and more particularly in the markets; this person entrusted with the *hisba* was called the *muhtasib*. There are two kinds of *hisba* treatise, the first being general dissertations on the virtue of *hisba* and the second manuals for the instruction of the *muhtasib*. Examples of the second type, which are our present concern, are known from the eleventh century onwards.[14] The *muhtasib* had many duties, which included the supervision of every trade in the market in order to exercise quality control and to guard against malpractices. In describing this function the *hisba* manuals tell us a good deal about the materials and methods of manufacture in the various trades. A second duty of the *muhtasib* was analogous to that of a modern clerk of works, in that he had to supervise the building of houses and the materials used in their construction. We can therefore learn from the manuals about the manufacture of bricks—burnt and unburnt—tiles, timber, etc. and how these materials were used in building construction. Information from these sources, however, although very valuable, is limited by the fact that the duties of the *muhtasib* were confined to the trades practised in the town. They tell us nothing about machines such as mills and water-raising devices, nor about large structures and public works.

The evidence from non-literary sources is very uneven. It is fortunate that we are fairly well-informed about fine technology from the *hiyal* treatises, because none of the small automata devices has survived. There are, however, many surviving examples of Islamic astrolabes and a few other types of astronomical instruments. In Fez, Morocco, there are the remains of two monumental water-clocks, built of masonry, which appear to have had water machinery very similar to that described in the clocks of Ridwān and al-Jazarī.[15] These remains are important, if only to demonstrate that the clocks described in the treatises were designs for actual clocks, rather than unrealized concepts. For despite the unmistakably practical intention of al-Jazarī's specifications, some writers have tried to show that his designs were simply paper ideas.

Many masonry structures built in medieval Islam still exist, and a number are still in use. The great mosques and palaces come immediately to mind, and there is plenty of literature available from which to arrive at a good understanding of the structural properties of these buildings. Some Muslim dams and bridges are still standing and substantial traces of others remain; a number of these can be examined in the works of modern writers.[16] Moreover, many of these structures can be visited without difficulty, although proper technical surveys would clearly be of more value than casual visits. In the field of hydraulic engineering, the construction of qanats using traditional techniques has been thoroughly studied and documented.[17] Medieval irrigation works have in many cases either been overlaid by new systems or have fallen into disrepair. Nevertheless, from the works that have survived, and from written records such as legal documents, histories and geographies, writers such as Thomas Glick have been able to present a coherent account of irrigation systems in certain communities.[18]

Our knowledge of heavy machinery comes mainly from archaeological sources, if this is taken to include the examination of traditional machines that are still in use. The *saqiya*, a chain-of-pots driven through a pair of gears by animal power, has been in continuous use for about two thousand years, and it is possible to verify, from scattered literary and iconographic evidence, that its design has not changed materially since the fifth century AD.[19] The noria, a water-driven wheel, has been in use for about the same period and, again, there is sufficient evidence to prove that its design has not been radically changed.[20] The case of water power, for grinding corn and for other industrial purposes, is rather different. Very few traditional water-mills are still in use, but in any case, as in Europe, the survival of an old mill does not necessarily tell us very much about its history. We cannot be sure, without further evidence, that there was a mill on the same site in medieval times, and in many cases milling installations were completely rebuilt on different lines in the eighteenth and nineteenth centuries. In the history of water power, certain important questions remain to be resolved. There are the questions of the origins, diffusion and relative importance of the three main types of wheel— overshot, undershot and horizontal. Of importance to economic and social historians is the extent to which water power was applied to industrial purposes. In an attempt to approach solutions to these problems we need to consult every type of evidence. Very little archaeological work has been done in this field; the remains of over thirty water-powered sugar-mills, dating to Ayyubid and Mamluk times, have been reported from excavations in Jordan.[21] Mills with horizontal wheels, some still working but most in disuse, are to be found in southern and eastern Europe, the Arab world and Iran. While the existence of these mills indicates a continuous tradition for their use, it does not tell us when the tradition began, nor does it permit us to locate the point of origin of this type of wheel in the geographical centre of the whole region.

The foregoing brief summary does not aim to be comprehensive, but it probably includes most of the important direct sources available to the student of Muslim engineering. At first glance it may seem that we are quite well supplied with information, but on closer inspection it will be seen that this is by no means the case. The best written sources are certainly those on fine technology—*hiyal*—but even in this field there is a barren period of about 350 years between the Banū Mūsà and al-Jazarī, and only a few brief works have survived from that time. In other fields, the deficiencies are more serious. We usually know what the finished constructions looked like and how they worked, from examinations of surviving machines and structures. On the other hand, none of the sources listed above tells us precisely what needs were served by particular constructions. In other words, how did technology serve the community? To attempt to answer this question, and to supplement our knowledge on purely technical matters, we must consult works of a general nature. Even so, we shall not find all the answers, partly because we cannot search everywhere and partly because some of the answers are nowhere to be found. One of the problems that bedevils the early history of water power, for example, both in Europe and in Asia, is that our sources very seldom tell us, in referring to a mill, the type of wheel which powered it. Indeed, they not infrequently fail to distinguish between the main sources of power—animals, water and wind. Even so, there is a good deal to learn by a careful reading of certain types of literature.

The surviving store of medieval Arabic literature, itself only part of the vast original production, is very large. Some, enough to fill the shelves of large libraries, has been edited and published, but even more is still in manuscript form, much of it uncatalogued. The task of cataloguing, editing, publishing and commenting upon the mass of manuscript material is an undertaking that would occupy the lifetimes of many scholars. Arabic writers of the Middle Ages were usually versatile, so that one expects to find references to many subjects in the works of a single scholar. From a technological point of view, for example, a close study of the works of the great scientist and polymath al-Bīrūnī would undoubtedly be very rewarding. His output, however, was voluminous and remains largely unpublished, so a study of this kind would take many years. One could suggest many other lines of research, all of them likely to be both rewarding and time-consuming. In the end, however, the best that the individual scholar can do is to dig into the strata which have a relatively high potential yield. For the historian of technology, the works of Muslim geographers and travellers have this potential.[22]

As mentioned at the beginning of this article, our main purpose is to indicate how the works of Muslim geographers may be used to increase our knowledge of Muslim engineering. Given that we are referring to a genre that comprises a large number of works, it would clearly be impossible in the space available to list all of them and review the merits and demerits of each one. I have therefore chosen to discuss the works of five writers, supplemented by citations from other works which I have had occasion to refer to in the course of my researches. The five are: al-Istakhrī, Ibn Hawqal and al-Muqaddasī—all geographers of the tenth century; Ibn Jubayr, the great Andalusian traveller of the twelfth century; and al-Idrīsī, a twelfth-century geographer, probably born in Morocco. The reasons behind this selection need not detain us long. It is not a 'representative sample' nor is it based entirely upon personal preference, although I must confess that I find all of these writers congenial. It is generally agreed that the tenth century witnessed the summit of achievement in Arabic geographical writing and that the three writers mentioned were the best of that school. Moreover, they wrote 'human geography' in the most flourishing period of Islamic civilization, and hence provide us with a mass of valuable information about economic and social conditions, not excluding technological matters, in the lands of the Muslim world. Ibn Jubayr was the first and one of the best of the Muslim writers of *Rihla* (*rihla*= travel, journey). His work has many similarities to the writings of the geographers, but devotes more space to personal experiences. Al-Idrīsī gives us more information than the other writers about life in North Africa and Spain, at a time of upheaval in the Iberian peninsula due to the successes of the Christian armies. All these works have been translated, in whole or in part, into modern European languages.

In his famous book *The Civilization of the Renaissance in Italy*, Jacob Burckhardt discusses how the Italians were the first Europeans to think of themselves as individuals:

> an *objective* treatment and consideration of the State and of all the things of this world became possible. The *subjective* side at the same time asserted itself with corresponding emphasis; man became a spiritual *individual*, and recognized himself as such. In the same way the Greek had once distinguished himself from the barbarian, and the Arab had felt himself an individual at a time when other Asiatics knew themselves only as members of a race.[23]

This way of thinking is very apparent in the works of the writers that we are discussing. They tell us about their own experiences, report conversations in the first person and, above all, consider nothing beneath their notice, provided it has some connection with human life. They can be scathing in their criticisms of governments, manners and morals, and living conditions. Only two factors detract somewhat from their value. All of them were members of the upper middle classes and their hosts and contacts were from the same class, so they seldom report directly on the lives of the rural or urban poor. Also, they were dependent upon patrons and tend, therefore, to paint a rosy picture of the lands under their control. The tenth-century geographers, for example, received benefits from the Samanid rulers of Khurasan, and they describe the province, especially the region of Bukhara and Samarkand, in glowing terms. This does not, however, lead to much distortion; it is generally accepted that the Samanids were enlightened rulers and that their lands prospered under their rule. In any case, although the writers may have exaggerated the wisdom of the rulers and the prosperity of the province somewhat, this did not prevent them from commenting objectively and sometimes adversely on the living conditions of the people.

For each province, the following information is given: first, a general description of the region, its extent, its boundaries and its principal towns and cities—in the manuscripts, maps were provided. There is a specialized vocabulary for each kind of locality—capital cities and seats of government, principal cities, military centres, towns, villages, suburbs, private and public lands, etc. Each city and its dependent towns, villages and countryside is then described in detail, including customs, religious divisions, dress and appearance of the inhabitants, and taxation. We are told about crops, manufacture, exports and imports, methods and materials of building construction, streets, roads and bridges. And always, water runs through the pages. For the smallest village we are invariably told where their drinking water came from—from wells, springs or from a stream. On a larger scale we have descriptions of irrigation systems, canals, qanats, dams, reservoirs and mills. For each province, the concluding section is concerned with communications, because these books were for the guidance of travellers. We are given routes, sometimes more than one, between the various towns in the province and the main routes to other provinces. This is done by stages of a day's journey, each with details of the shelter, provisions and water available at each staging point. The above summary applies to the geographers' books. Much of the same information is found in Ibn Jubayr's work, but there is much less in the way of topographical information and, of course, more about his own experiences.

Very little is known about the life of al-Istakhrī, the earliest of our writers, whose full name was Abu Ishāq Ibrāhīm b. Muhammad. His name implies that he came from Istakhr, the Arabic name for Persepolis in the Iranian province of Fars, and he may have lived for some time in Baghdad. He almost certainly visited Arabia, Iraq, Khuzistan, Daylam (the province south-west of the Caspian) and Transoxiana. His work was written towards the end of the first half of the tenth century. The best edition is that of al-Hīnī (to which the page references apply) and there are partial translations in German and Italian.[24] The work is entitled *Book of the Routes and the Lands* (*Kitāb al-masālik wa'l-mamālik*).

Ibn Hawqal, Abu'l-Qāsim b. 'Alī al-Nasibī was a native of Nisibis in Upper Mesopotamia. Between 947 and 973 he travelled widely in the

Muslim world—to Spain, North Africa, Egypt, Iraq, Iran and Central Asia. His book, *Configuration of the Earth* (*Kitāb Sūrat al-Ard*), appeared in its final, definitive version about 988. As a young man, Ibn Hawqal had met al-Istakhrī and became in some ways his follower; many of the reports in his work are word-for-word copies of passages in al-Istakhrī's book. Ibn Hawqal, however, is very far from being a mere plagiarist. Not only was he able to make first-hand reports on places that al-Istakhrī had not visited, but his work has a breadth and a personality incomparably larger than that of his predecessor. Of all the Arab geographers, his work is probably of the greatest value for modern historians. To quote André Miquel:

> His constant care to depict a region precisely in the state and at the date that he himself had seen it, and occasional references to the distant or more recent past, give to his text, besides a vividness of description and even a depth of feeling which sometimes appear, undoubted value for the historian. This is particularly true of the notes on economic matters, which form a complete break with convention; for one thing, Ibn Hawkal is much less interested in rare or precious products than in the basic agricultural and manufactured products, and secondly he was able to study on the spot a given economic situation in relation to a particular period or with reference to an implicit norm. He was the only Arab geographer who really sketched a vivid picture of production.[25]

The Arabic text was edited by J.H. Kramers and there is a full French translation, based on Kramers' text, by G. Wiet (notes refer to pages in the Arabic text).[26]

Al-Muqaddasī, Abu 'Abd Allah Muhammad b. Ahmad (d. 1000) was a contemporary of Ibn Hawqal. His book is entitled *The Best Division for the Understanding of the Provinces* (*Ahsan al-taqāsīm lī ma 'arifat al-aqālīm*), completed between the years 985 and 990 after its author had spent many years travelling throughout the Muslim world.

> He rightly claims to have put Arab geography on a new foundation and given it a new meaning and a wider scope. Since he considered the subject useful to many sections of society, as also to the followers of many vocations, he widened its scope, including in it a variety of subjects ranging from physical features of the *iklīm* (region) under discussion to mines, languages and races of the peoples, customs and habits, religions and sects, character, weights and measures and the territorial divisions, routes and distances. He believed that it was not a science that was acquired through conjecture (*kiyās*), but through direct observation and first-hand information.[27]

Much of his work makes very lively reading. He expresses his opinions of people and places pungently, and his descriptions are animated and evocative. He was a native of Palestine, and often compares foreign towns to towns in Palestine, nearly always to the detriment of the former. The

Arabic text is edited by de Goeje and there is a partial French translation (references to the Arabic text).[28]

Ibn Jubayr, Abu'l-Husayn Muhammad b. Ahmad, was born in Valencia in 1145 and died in Alexandria in 1217. He made three journeys, but his fame rests on his day-by-day account of the first of these. He sailed from Tarifa in Morocco early in 1183 in a Genoese ship which took a month to reach Alexandria by way of Sardinia, Sicily and Crete. He then travelled to Mecca, where he spent nine months before continuing his journey to Medina, Iraq, Upper Mesopotamia, Aleppo, Damascus and Acre (then in the hands of the Crusaders). He took ship from there, again in a Genoese vessel, bound for Sicily and narrowly escaped with his life in a shipwreck in the straits of Messina. Re-embarking at Trapani, he arrived at Cartagena in March 1185.[29] He gives us an abundance of information about the peoples among whom he stayed and about navigation, buildings and communications. The standard edition is that of Wright, emended by de Goeje, and there are full translations in English, French and Italian.[30] (References to the Arabic text.)

Al-Idrīsī, Abu 'Abd Allah Muhammad b. Muhammad completed his geography in 1154 at the court of Roger II of Sicily, on whose orders the book was written. For this reason it is usually called simply *The Book of Roger*. Biographical references to him are rare, probably because the Arab biographers considered him a renegade for having lived at the court of a Christian king and written in praise of him. He was born about 1100, perhaps in Ceuta, and may have studied at Cordoba before settling in Sicily, where he died about 1165. His work is a straightforward geography and contains a good deal of information about economic affairs in Spain and North Africa. Part of the book was edited by R. Dozy and M.J. de Goeje in 1866, with a French translation, and I have used this work in preparing this article (references are given to both the original and the translation, with the Arabic first, for example 187/228). A full critical edition of the *Book of Roger* is being undertaken by an international group of scholars, under the auspices of the Instituto Italiano per il Medio e l'Estremo Oriente.[31]

One of the main concerns of these writers is to depict human society in each part of a given province within its urban and rural environment. The second concern is with commerce—with the production of goods and materials and their interchange, inside the Muslim world and with foreign countries. They do not, in general, seem to have been very interested in the daily lives of the inhabitants. Descriptions of cities and towns are usually fairly brief and are confined to details of layouts, with the locations of the principal buildings, markets, main streets and gates. Building materials are mentioned in very broad terms, for example 'their houses are built of wood/burnt bricks/stone', and only important buildings such as large mosques are described in detail. On the other hand, there is a wealth of information about agriculture: the various types of crops are listed, we are told whether cultivation was by irrigation or rainfall and about methods of processing—grist milling, drying, pickling and salting.

Often this processing was done in order to prepare the foodstuffs for export. We are seldom told, however, precisely how this processing was done but simply that it *was* done. Textile materials, of course, also came from the fields, and our writers mention a bewildering variety of fabrics made from wool, cotton, flax and silk. These fabrics, which were woven in state-operated workshops called *tirāz*, were the most important commercial commodities in the Muslim world.[32]

For the most part, our information about technology is a by-product of our writers' concern with agricultural production and their pre-occupation with water. So there is a relative abundance of data about the hydraulic works necessary to impound water and conduct it to the fields and urban communities. At the other end of the cycle of production, we can learn a good deal about the scale of corn-milling operations. Frequently, the reports are of a general nature, describing a river and the various irrigation canals drawn from it and the lands watered by them. Such, for example, is al-Muqaddasī's description of the canal system in Transoxiana (p. 280ff.). The irrigation system in the region of Marw, on the River Murghāb in Khurasan, was extensive and highly organized. Ibn Hawqal tells us that the director of the system had more power than the prefect of the city and controlled a labour force of 10,000 paid employees who carried out maintenance work and supervised the allocation of the water (pp. 635-6). Al-Muqaddasī adds that included in the labour force was a team of 400 divers who kept the dams and dykes under 24-hour observation. Each man had to keep with him a supply of timber and brushwood for carrying out emergency repairs and if they had to enter the water when it was very cold they coated their bodies with wax (pp. 330-1). All the water in Nishapur, on the other hand, was supplied by qanats, both for the town supply and for irrigation (Istakhrī p. 145). The same was true for Tangier in Morocco, where the main feeder qanat was many miles in length (Ibn Hawqal p. 79). A number of new cities were founded by the Muslims, and sometimes completely new irrigation systems had to be constructed as part of the cities' infrastructure. The fullest description of the excavation of such a system, for Basra in Iraq, occurs in the work of the historian al-Balādhurī (d. *c.* 892). The city was founded as a temporary military encampment in 638, but the two main feeder canals from the Shatt al-ʿArab were not completed until about 660. Thereafter, the excavation of new canals, which al-Balādhurī lists in detail with the names of their constructors, kept pace with the growth of Basra from a settlement into a large city.[33] About 150 years later, even though Basra was somewhat in decline, al-Istakhrī was amazed to see the vast network of canals (p. 57).

Dams were used for various purposes and they are sometimes described in some detail. Al-Muqaddasī describes two dams in Egypt, made from earth and esparto, which were breached at the time of the Nile flood to allow the water to flow on to the fields (p. 206). Much more impressive are his descriptions of two large dams, both of which provided hydropower as well as irrigation water. One of these was built in pre-Islamic times at Ahwāz in Khuzistan (p. 411). The other was on the River Kūr, in the central Iranian province of Fars; it was built in the tenth century by the Buwayid Amir ʿAdud al-Dawla:

'Adud al-Dawla closed the river between Shiraz and Istakhr by a great wall, strengthened with lead. And the water behind it rose and formed a lake. Upon it on the two sides were ten water-wheels, like those we mentioned in Khuzistan, and below each wheel was a mill, and it is today one of the wonders of Fars. Then he built a city. The water flowed through channels and irrigated 300 villages (p. 444).

This dam, known as the Band-i Amir, still exists today. Al-Idrīsī describes a dam at Cordoba in Spain: 'across the river, below the bridge, was a dam made of Quptiyya stone, with large marble pillars. In the dam were three millhouses and in each house were four mills' (p. 212/262). Until recently its three millhouses still functioned, but much changed from their original form.[34]

These writers convey a very positive attitude towards the use of water power. The Tigris at its source, says al-Muqaddasī (p. 136), would turn only one mill, and al-Istakhrī, looking at a fast-flowing stream in the Iranian province of Kirman, estimates that it would turn twenty mills (p. 166). There are frequent references to mills all over the Muslim world, from the Iberian peninsula to Central Asia, and near large cities there were often dozens of mills constituting, as it were, factory milling complexes. It could scarcely have been otherwise. At the height of its prosperity, Baghdad had a population of about 1·5 million and Fustāt (Old Cairo) and Cordoba, though smaller, were still very populous cities. The populations were agriculturally non-productive and had to be fed, clothed and housed, and their merchants supplied with the materials of commerce. One example of large-scale milling, from Ibn Hawqal, will have to suffice. It refers to Upper Mesopotamia, which was the granary for Baghdad:

The ship-mills on the Tigris at Mosul have no equal anywhere, because they are in very fast current, moored to the bank by iron chains. Each has four stones and each pair of stones grinds in the day and night 50 donkey-loads. They are made of wood—sometimes of teak (p. 219).

If we take a donkey-load as 100 kg, then the output of one of these mills in 24 hours was 10 tonnes. Long after the time of Ibn Hawqal, Upper Mesopotamia continued as a large supplier of flour to Iraq. About 1184, Ibn Jubayr saw the ship-mills across the River Khabur, 'forming, as it were, a dam' (p. 243).

Ship-mills were one way of increasing the output of mills, by taking advantage of the high velocity of the water in midstream. For the same reason, water-wheels were installed on the piers of bridges (al-Muqaddasī p. 312). The use of dams to increase the head of water above the mills has already been mentioned. The power of the tides was also harnessed.

Al-Muqaddasī mentions tidal mills in the Basra area: 'The ebb-tide is also useful for operating the mills, because they are at the mouths of the rivers, and when the water comes out it turns them' (pp. 124–5). The quality of millstones has always been important. They must be hard but of homogeneous texture, so that pieces of grit do not become detached and get mixed with the flour. The stone from certain localities was therefore particularly prized for milling purposes and there are a number of references to suitable kinds of stone. In the area of Majjāna, in modern Tunisia, for example: 'They cut millstones from the nearby mountains, which are excellent and good for grinding; they can last a man's lifetime without dressing or other treatment, due to their solidity and the fineness of their grains' (al-Idrīsī pp. 118/138).[35]

There is therefore a good deal of information about mills and their importance in the economic life of the Muslim world. We do not learn much, however, about the actual construction of the machines—Ibn Hawqal's description of ship-mills, quoted above, is one of the few reports that gives any constructional details. We know from the *hiyal* treatises and elsewhere, however, that all the basic types of water-wheels were in use in the Muslim world, and we can infer that the type of wheel used in a given locality was that which best suited the local conditions. Nor do we find much information about industrial uses for water power, apart from corn milling. There is an indication in al-Muqaddasī that water power was utilized for the fulling of cloth (p. 409), but we must use other sources for information about industrial milling.[36]

The only references to windmills are to the machines with vertical axles which were built in the province of Sijistan or Seistan (the western part of modern Afghanistan). Al-Istakhrī was the first to mention them (p. 140), but for a full description of this type of machine, which were still in use until recently, we have nothing earlier than a passage by the Syrian geographer al-Dimashqī, in a book written about 1271. He tells us that the mills were supported on substructures built for the purpose, or on the towers of castles or on the top of hills. They consisted of an upper chamber in which the millstones were housed and a lower one for the rotor. The axle was vertical and it carried twelve or six arms covered with a double skin of fabric. The walls of the lower chamber were pierced with funnel-shaped ducts, with the narrower end towards the interior in order to increase the speed of the wind when it flowed on to the sails.[37]

There is abundant evidence for the widespread use of water-raising machines in the Muslim world and they are often mentioned by our writers, although most references simply state that saqiyas or norias were used in a given area. The first report for the construction of norias in Islam comes from al-Balādhurī, who tells us that in the second half of the seventh century (we can date it no closer) a certain Bilāl excavated a canal in the Basra area and erected norias along its bank.[38] Al-Muqaddasī was clearly impressed with the large norias on the river at Ahwāz in Khuzistan (p. 411) and al-Idrīsī describes a hydraulic installation at Toledo in which a noria lifted water through a height of 50 cubits to an aqueduct that led into the city (pp. 187/288). In the town of Ra's al-'Ayn in Upper Meso-potamia, norias raised water from the River Khabur on to the gardens (Ibn Jubayr p. 243).

When we leave the field of hydraulic engineering, we find that reports are less frequent and generally briefer. An exception, as mentioned earlier, is the case of large buildings such as great mosques, for which the descrip-tions are often sufficiently detailed to be of value to students of architec-ture and building construction. On the other hand, many of these build-ings still exist and can be examined *in situ* or through the works of architectural historians.

Another type of large structure that often invited comment was the arched bridge, built of masonry or occasionally of burnt bricks. Many of those described were of pre-Islamic origin—Roman or Sasānid. Al-Idrīsī, for example, describes the many-arched Roman bridge over the Guadalquivir at Cordoba (pp. 212/262). One bridge, which still exists, impressed all the Muslim writers who saw it. This is the bridge at Sanja in Upper Meso-potamia, a single arch of dressed masonry with a span of 112 feet built by the Emperor Vespasian about AD 70 (see, for example, Ibn Hawqal p. 123). A remarkable Muslim bridge, built over the River Tab in central Iran, is described, among others, by al-Istakhrī. He says that it was built by an Iranian, physician to the Umayyad governor al-Hajjāj (661–714). It was a single arch of 80 paces in span, and so high that a man on a camel with a long standard in his raised hand could pass beneath it (p. 91). Al-Qazwīnī, in his geography completed in 1276, has left us a graphic account of the rebuilding of a ruined Sasānid bridge at the town of Idhaj in Khuzistan. It was built by a minister of the Buwayid Amir al-Hasan, who conscripted craftsmen from Idhaj and Isfahan. The bridge was 150 cubits in height and consisted of a single arch supported on tapering masonry piers strengthened with lead dowels and iron clamps. The slag from iron workings was used to fill the space between the arch and the roadway.[39] Bridges of boats were common and are frequently mentioned, although seldom with any details of their construction. Ibn Jubayr, however, describing a bridge of boats over the Euphrates at Hilla, tells us that it had chains on either side to which the boats were attached. These chains, 'like twisted rods', were secured to wooden anchorages on the banks (p. 213).

References to fine technology are very rare and add little to the infor-mation in the *hiyal* treatises. Al-Muqaddasī remarked upon the beauty of the fountains in Damascus (p. 157) and about two centuries later, in the same city, Ibn Jubayr saw a fountain consisting of a single jet surrounded by small pipes that threw the water up like the branches of a tree (p. 269). Al-Muqaddasī refers to a portico by the Jayrun Gate in Damascus as the 'Gate of the Hours', indicating that a clock had traditionally been located at this spot, where Ridwān's father was to build his clock in the twelfth century (p. 158). This clock was seen and described by Ibn Jubayr; his description, though quite brief, tallies with the details given by Ridwān in his treatise (Ibn Jubayr p. 270 ff).

The foregoing discussion and extracts, it is hoped, indicate how the works of Arab geographers can assist the historian of technology. Quite apart from their value in this respect, however, writers such as Ibn Hawqal and al-Muqaddasī paint a vivid picture of the Muslim world at the peak of its prosperity. While not free from blemishes and prejudices, their works depict that society as they saw it, with its merits and its defects. One thing is certain: it was a society that made extensive use of technology to support large populations and a thriving commerce.

Notes

(*EI* is used as the abbreviation for *Encyclopaedia of Islam*, 2nd edn, Brill, Leiden, 4 vols. to date, continuing)

1. Donald R. Hill, 'Hiyal' in *EI*, suppl. to vols I to III, pp. 371–4.

2. Banū Mūsà, *The Book of Ingenious Devices*, trans. and annotated by Donald R. Hill, Reidel, Dordrecht, 1979. *Kitāb al-Hiyal*, Arabic edn of the Banū Mūsà's work, ed. A.Y. Hassan, Institute for the History of Arabic Science, Aleppo University, 1981. Donald R. Hill, 'The Banū Mūsà and their Book of Ingenious Devices' in *History of Technology*, Vol. 2, 1977, pp. 39–76.

3. Al-Khuwārizmī, Abu ʿAbd Allah Muhammad, *Liber Mafātīh al-Olūm*, ed. G. van Vloten, Arabic text with Latin critical apparatus, Brill, Leiden, 1895.

4. Donald R. Hill, *Arabic Water-clocks*, Institute for the History of Arabic Science, Aleppo University, 1981, pp. 36–46.

5. *Ibid.*, pp. 47–68.

6. *Ibid.*, pp. 69–88.

7. Al-Jazarī, Ibn al-Razzāz, *The Book of Knowledge of Ingenious Mechanical Devices*, trans. and annotated by Donald R. Hill, Reidel, Dordrecht, 1974. *A Compendium on the Theory and Practice of the Mechanical Arts*, Arabic text of al-Jazarī's work, ed. A.Y. Hassan, Institute for the History of Arabic Science, Aleppo University, 1979.

8. Philo of Byzantium, 'Le Livre des Appareils Pneumatiques et des Machines Hydrauliques par Philon de Byzance', Arabic text edn with French trans. by Carra de Vaux, *Paris Académie des Inscriptions et Belles Lettres*, Vol. 38, Part 1, 1903.

9. *On the Construction of Water-clocks*, trans. and annotated by Donald R. Hill, Occasional Paper No. 4, Turner & Devereux, London, 1976.

10. E. Wiedemann, *Aufsätze sur Arabischen Wissenschaftsgeschichte*, 2 vols., Olms, Hildesheim, vol. 2, 1970, pp. 50–6.

11. There are, for example, a number of manuscripts containing a chapter called 'Wheels that move by themselves'. Despite the perpetual motion element in these devices, they would repay close study. See Thorkild Schiøler, *Roman and Islamic Water-Lifting Wheels*, Odense University Press, 1973, pp. 72–8.

12. al-Karajī, *Inbāt al-miyāh al-khafiyya*, Hyderabad, 1945.

13. C. Cahen, 'Le Service de l'irrigation en Iraq au début du XIᵉ siècle' in *Bulletin d'études orientales*, Vol. 13, 1949–51, pp. 117–43.

14. C. Cahen and M. Talbi, 'Hisba' in *EI*, vol. III, 1971, pp. 485–8.

15. Derek de Solla Price, 'Mechanical Water-clocks in the 14th century in Fez, Morocco' in *Ithaca*, Hermann, Paris, 1962, Vol. 26, pp. VIII–IX.

16. For example, Norman Smith, *A History of Dams*, Peter Davies, London, 1971.

17. H. Goblot, *Les Qanats; une Technique d'Acquisition de l'Eau*, Mouton Éditeur, Paris, 1979; and A.K.S. Lambton, 'Kanāt' in *EI*, vol. IV, 1978, pp. 528-32.

18. Thomas F. Glick, *Irrigation and Society in Medieval Valencia*, Harvard University Press, 1970.

19. Thorkild Schiøler, as note 11.

20. Donald R. Hill, *A History of Engineering in Classical and Medieval Times*, Croom Helm, 1984, pp. 139-42.

21. S. Hamarneh, 'Sugar-cane Plantation and Industry under the Arab Muslims during the Middle Ages' in *Proceedings of the First International Symposium for the History of Arabic Science*, Aleppo University, 1976, p. 221.

22. Two modern works on the early Muslim geographers are worthy of mention. The first of these is by André Miquel, *La géographie humaine du monde musulman jusqu'au milieu du XIᵉ siècle*, Paris, 1967. This is a detailed study of geographical writings in Islam, from the rise of Muslim literature until about 1050. The development of scientific geography is discussed and the achievements of individual geographers are assessed. The second work is by Guy le Strange, *The Lands of the Eastern Caliphate*, Frank Cass, London 1905, third impression 1966. From the works of a number of Muslim geographers, the author reconstructed the physical geography of medieval Islam, from Iraq eastwards. There are maps for each province and a good index.

23. Jacob Burckhardt, *The Civilization of the Renaissance in Italy*, Phaidon Press, Oxford and London, 1945, p. 81.

24. Arabic edition by M.G. ʿAbd al-ʿĀl al-Hīnī, Cairo, 1961; German trans. *Das Buch der Länder*, A.D. Mordtmann, Hamburg, 1845, partial Italian trans. by A. Madini, Milan, 1842; see A. Miquel, 'al-Istakhri' in *EI*, Vol. IV, 1978, pp. 222-3.

25. A. Miquel, 'Ibn Hawkal' in *EI*, Vol. III, 1971, pp. 786-8.

26. Arabic edition by J.H. Kramers, 2nd edn of *Biblioteca Geographorum Arabicorum* (BGA), Brill, Leiden, 1938; French trans. *Configuration de la Terre*, by J.H. Kramers and G. Wiet, 2 vols., Beirut/Paris, Vol. 1, 1964, Vol. 2, 1965.

27. S. Maqbul Ahmad, 'Djugrafiyya' in *EI*, Vol. II, 1965, p. 582.

28. Arabic edn by M.J. de Goeje, vol. 3 of BGA, Brill, Leiden, 1906; partial French trans. by A. Miquel, *La meillure répartition pour la connaissance des provinces*, Damascus, 1963.

29. C. Pellat, 'Ibn Djubayr' in *EI*, Vol. III, 1971, p. 755.

30. Arabic edn by M.J. de Goeje, Leiden, 1907, from the 1852 edn by M. Wright; Italian trans. *Viaggio in Ispagna, Sicilia, etc.* by C. Schiaparelli, Rome, 1906; English, *The travels of Ibn Jubayr*, by C. Broadhurst, London, 1952; in French by M. Gaudefroy-Demombynes, 3 vols, Paris, 1949-56.

31. Arabic with French trans. of part of Idrīsī's work: *Description de l'Afrique et de l'Espagne*, by R. Dozy and M.J. de Goeje, Brill, Leiden, 1866; see also G. Oman, 'al-Idrīsī' in *EI*, Vol. III, 1971, pp. 1032-5.

32. R.B. Serjeant, *Islamic Textiles—Material for a History up to the Mongol Conquest*, Beirut, 1972.

33. Al-Balādhurī, *Kitāb futūh al-buldān* (*Conquests of the Lands*), ed. M.J. de Goeje, Brill, Leiden, 1866, pp. 345-71.

34. Norman Smith, *Man and Water*, Peter Davies, London, 1976, p. 143.

35. Donald R. Hill, as note 20, p. 166.

36. For example, al-Bīrūnī in his work on minerology, *Kitāb al-jamāhir etc.*, ed. F. Krenkow, Hyderabad, Deccan, 1936, pp. 233-4, tells us that water-driven trip-hammers were used for pounding gold ores 'as is the case in Samarkand with the pounding of flax for paper'.

37. Al-Dimashqī, *Cosmographie*, Arabic text edn A.F. Mehren, St Petersburg, 1866, p. 182.

38. Al-Balādhurī, as note 33, p. 363.

39. Al-Qazwīnī, *Athār al-bilād*, Beirut, 1960, p. 303.

The Second-Century Romano-British Watermill at Ickham, Kent

ROBERT J. SPAIN

Introduction

At the beginning of the twentieth century, no watermill sites were known or recognized in Roman Britain, but in the next quarter century their existence was confirmed by the identification of three sites associated with Hadrian's Wall. The first site to be discovered and recognized as a water-mill, and probably the best known of all Romano-British watermills, was Haltwhistle Burn Head which was found and interpreted by F.G. Simp-son[1] in 1907–8. He found the remains of a stone-walled building of rec-tangular plan, beside a watercourse, in the bottom of which were several millstones and querns. No trace of the machinery remained but the waterwheel was clearly of the undershot type, approximately 3·6 m in diameter and 0·35 m wide.

During 1923 and 1924, R.C. Shaw, excavating at the Roman bridge where the Wall passes over the River Irthing near Willowford, found stone-lined sluices and millstone fragments.[2] This site is now accepted as probably that of a watermill, and so too is the tower at the eastern abutment of Chollerford Bridge near Chesters fort. This was excavated in the 1860s by J. Clayton, who found a stone-lined watercourse passing through the base of the tower together with millstone fragments, but apparently he failed to identify it as a watermill.[3]

After mid-century, increasing activity in British archaeology combined with a greater awareness of watermills produced several more possible sites. Numerous Romano-British artefacts including millstones, pottery and tiles were found on the site of a modern watermill at Spring Valley Mill near Ardleigh, Essex.[4] Close by Fullerton Villa in Hampshire, D.B. Whitehouse found a flinty rubble platform with traces of several post holes beside an ancient stream.[5] In the adjacent bifurcated stream, the remains of two millstones were found. Unfortunately, time did not permit a full excavation of this threatened site but it is now accepted as a possible watermill. At Nettleton in Wiltshire, W.J. Wedlake has found the remains of a most unusual watermill, apparently of Roman date.[6] Erosion of the banks of the Broadmead Brook revealed a narrow waterwheel emplace-ment and channel formed of dressed limestone with a splayed water in-take, an inclined sluice-gate and curved breast showing that the wheel was 2·6 m in diameter. Near Leeds village in Kent a millstone fragment was found with Romano-British sherds and a brooch on the banks of a stream which had incised itself deep into the surrounding farmland.[7] No positive building remains were located and the sherds were very water-worn suggesting that the evidence had been eroded and scattered well down the valley. Several other places have been suggested as possible watermill sites but none have yielded positive evidence of a mill structure or waterwheel.[8]

During the summer of 1974, however, two most interesting Romano-British watermills were unearthed at Ickham near Canterbury. The second mill to be discovered, which proved to be the oldest, was excavated by the Ashford Archaeological Society under the direction of J.V. Brad-shaw. Further work on the site revealed the remains of another mill a short distance upstream which was of fourth century AD construction. This was excavated by the Department of Environment under the direction of C.J. Young.[9]

The Site and its Ground Conditions

The site of the watermills appears to be part of a large area of Roman occupation (see Fig. 1). From 1958 onwards, Romano-British finds were made by many people[10] and burials, walls and floors, pottery, drainage ditches and coins were found. In the early 1970s, extraction of river gravel by the wet-pit method using floating dredgers brought to light more evidence. Large parts of the settlement were probably destroyed before a salvage and rescue operation was organized in 1974. The remains of a road were found crossing the area at several places and, most significantly, an old course of the Little Stour River was found passing through two timber structures which were subsequently recognized as watermills. At each site the remains of numerous post ends were found well preserved by the waterlogged soil. The high concentration of posts at the smaller and older watermill, which appears to have been entirely earth-fast, has facili-tated a reconstruction of the building and its machinery, and that is the subject of this paper.

First of all, however, some consideration ought to be given to the nature of the ground. The present site is low-lying and the surrounding fields are intersected by numerous ditches and branches of the Little Stour. In recent times the area between Stodmarsh and Grove Ferry, immediately north of the mill site, has suffered from subsidence caused by the mining operations of the nearby Chislet Colliery. Furthermore the extraction of river gravel by wet pits has created numerous flooded pits with charac-teristic bank vegetation of reeds. One canalized branch of the river passes some 90 m to the north-west of the site to join the present tidal head amid marshes a little over a kilometre to the north.

Aerial survey has revealed that an old course of the river ran through the mill site and alongside, on the north bank, ran a Roman road. This previously unrecorded road ran from Great Wenderton, on the east bank of the Wingham River at O.S. reference TR 2355 5923, across Horse

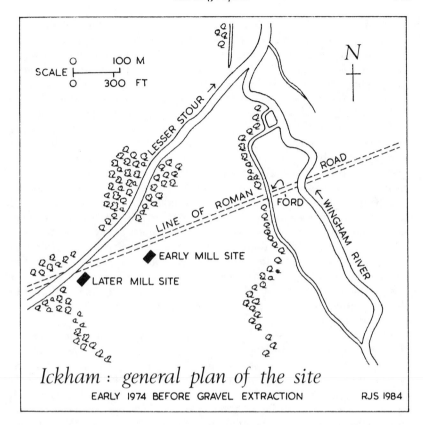

SCALE

0 ———— 100 M
0 ———— 300 FT

N

Ickham : general plan of the site

EARLY 1974 BEFORE GRAVEL EXTRACTION   RJS 1984

Figure 1.

Marsh, and apparently terminated at Quaives, east of Wickhambreux village at O.S. reference TR 2245 5880. The alignment of this road is 248·5° from true north. This road may have been connected with communication between the Roman station at Richborough (*RVTVPIAE*) and Canterbury (*DVROVERNVM CANTIACORVM*) but there is no evidence to support this suggestion and the purpose of the road may have been solely for local needs. This theory accords with the unpublished findings of J.D. Ogilvie's excavations of 1958 and 1959.[11]

The construction of this road is worthy of closer attention for it gives us an indication of the ground conditions during the period of Roman occupation. Roman roads varied considerably in their construction, for example in the number of material layers, thicknesses and widths, but they invariably had foundations of local materials such as sand, stone, gravel or iron slag.[12] However, at Ickham the road passing the mill is unusual for, while traditional foundations occurred in the approaches to the mill from the west, some 180 m beyond the mill a section was revealed in which the metalled road lay on pegged brushwood. Further on some

270 m from the mill, this road crossed a contemporary ford of flints and where it approached the river crossing evidence was found of the road lying on logs pegged down with short stakes, suggesting that the increasing instability of the ground had necessitated another change in the foundation technique. So we may conclude that the mill was sited close to, perhaps on the very edge of, the marshes bordering the Wanstsum.

The remains of most of the larger posts were found to have their buried ends roughly hewn by adze or axe, but obtuse and not sharp, indicating that they had been dug in and not driven as some of the smaller posts and stakes probably were. It has been known for a stone to be placed below a post to prevent it sinking in soft or waterlogged sub-soils but no such stones were discovered at Ickham.[13] No iron-shod posts were found, as have been met with in some Roman piling, nor was there evidence of their ends having been previously charred to prevent decay.[14] Having been dug in they would not have been placed much below the water table and so the building conditions may have been superior to the excavating conditions, when the water table was well above the post ends.[15] The relative movements of land and water levels that have occurred since their emplacement have caused this. With the very limited time available to the excavators, it was not possible to identify and define the pits dug for the erection of the larger posts.

The ground itself was brickearth and not far below gravel, from an old river terrace. The exceptional state of preservation of the foundation timbers was due to the considerable depth of peat on the site. However, during the life of the mill the decay of timber would have been most pronounced at ground level, especially in posts which contained a high proportion of sapwood. Below ground level, where the nature of the soil has a great effect on the rate of decay, the heavy, badly drained conditions probably inhibited fungal growth because of insufficient aeration. Why so low-lying and semi-marshy a site was selected for a mill is not clear unless it was to take advantage of water carriage. The task of construction would have been easier on higher, or firmer, ground where a raft of oak shingles and/or ground beams could have been used or the timber frame carried upon stone sill-walls.

The evidence from this site does not indicate whether the mill was built on an existing stream or if a new watercourse was cut. Due to the water-logged nature of the ground it was not possible to determine the route of the watercourses in the environs of the mill. However, information was obtained from the second and later mill site built some 140 m upstream at a point where a bifurcation had been made. Both the millstream and bypass channels were found during the excavation and evidence of their sections, particularly widths, gives us a reliable impression of the water flow rate that probably existed. It seems likely, in view of the amount of revetment work in the millstream of the second site, that this was an artefact and that the bypass was the natural course. A comparison between this original course upstream and the headrace entering the lower mill suggests that the lower mill was not built over the original stream for the width of the headrace in the lower mill is much less than the upstream bypass course. But this is to be expected because there were advantages in not building beside an existing stream. This is particularly so when building a mill to straddle the water because the new millstream can be dug and brought to the mill on whatever line is most suitable and the wheel pit can be constructed to conform precisely with the design of the wheel. If construction took place over an existing channel, the banks would almost certainly require revetment and consolidation, especially where the wheel was to work.

Although the mill must have stood very close to estuarine land, the tidal water would not have been permitted to create backwater and thereby affect the working of the wheel. The bed of the headrace was probably sloped to give velocity to the water and the tail race similarly graded so that in times of spate the water readily cleared the wheel, but it has not been possible to confirm this. From the evidence found, there can be little doubt that the waterwheel was undershot and this point is discussed later.

A Note on Terminology

When discussing watermills it is important to employ, where appropriate, the correct terminology as applied to the machinery, waterways and supporting structures. In comparatively recent times the oral traditions of wind and water milling have given us a language of corn milling which has evolved over centuries and is peculiar to that industry. Although it suffers from local variations, as does any long-established industrial language, many of the words and terms have a common form. The application of modern terms to a watermill which worked one and three quarter millenia ago might be questioned, but it is enough that in this ancient mill there were many mechanical and structural elements having similar, and in some cases identical, functions to those found in existing watermills. We must remember that this style of watermill, with its vertical wheel, was new to the ancient world. The Augustan engineer Vitruvius, writing in approximately 25 BC, describes them in his work *De Architectura*.[16] His writing implies that these mills were fairly rare. Where his mechanical terms are identifiable, they have been given in the first instance.

With respect to the orientation of axes, the terms longitudinal and transverse are applied relative to the axis of the waterwheel shaft. An alternative would be to use the watercourse as the datum but this has the disadvantage of confusion when the axis of the head and tail race sections of the stream deviate from that of the wheel rotation, or are unknown. Both these disadvantages arise at this site. However, the main reason why the wheel-shaft axis has been taken as datum is because the analysis of evidence is primarily concerned with the transmission of power, not its generation.

Structural Analysis—Support of the Machinery and Millstones

In modern timber-framed watermills the structures for supporting the millstones are usually integrated with the building frame. However, we must not be influenced by medieval or modern designs because they occur much later in the evolution of watermill structures and we should therefore approach this analysis with an open mind.

A preliminary examination of the foundation plan, see Fig. 2, reveals that there are several areas where the posts appear to be grouped together. The groups are identified as follows: posts 1–3; posts 4–8; posts 14–17 and posts 18–22. Nearly all of them are of quadrilateral section and in each group the axes of the individual posts are parallel to each other. Outstanding examples of this symmetry are groups 4–8 and 14–17; it can also be seen, although less obviously, in group 18–20, while in group 1–3 the axis of round posts 2 and 3 taken together is parallel to the minor[17] axis of post 1. Of greater importance is the inter-group relationship whose principal feature is that the major axes of groups 1–3, 18–22 and 14–17 are parallel to each other. The major axis of group 4–8 is not parallel to the other three groups but its conformation is outstanding and its density the greatest on the site.

In addition to the four post groups so far identified there is another group which shows signs of having some relationship. These are posts 9–13, 26 and 27 which have common minor axes all parallel to the groups' major axis—but they have a linear grouping, unlike the other four which are clusters.

Finally, all five groups may be related by the fact that the major axis of group (9–13, 26, 27) is closely parallel to the combined major axis of groups [(1–3) and (18–22)] and to the common axis of groups [(4–8) and (14–17)]. An alternative view of their relationship can be seen by noting that the centre of area of each of the four cluster groups lies on transverse axes parallel to the axis of the linear group (9–13, 26, 27).

To analyse the structure it is necessary to recognize the support requirements of an early vertical-wheeled watermill. In addition, consideration must be given to accessibility for maintenance and operation. There are three basic support functions: (i) the horizontal shaft carrying the waterwheel and driver gear (*tympanum*);[18] (ii) support for the vertical spindle[19] with the driven gear mounted on it and the upper or runner millstone which the spindle drove; and (iii) support for the bed-stone or lower millstone, the grain bin or hopper (*infundibulum*) and feeding mechanism together with the floor on which the millstones sat.

Water travelling in the headrace entered the mill between posts 23 and 24 to drive the undershot waterwheel straight ahead. It is quite likely that the headrace was graded just prior to the wheel to promote greater water velocity. The width of the wheel was limited by the distance between posts 13 and 15 which, allowing for satisfactory clearances, could not have been much more than 0·565 m. The diameter of the wheel is not so readily

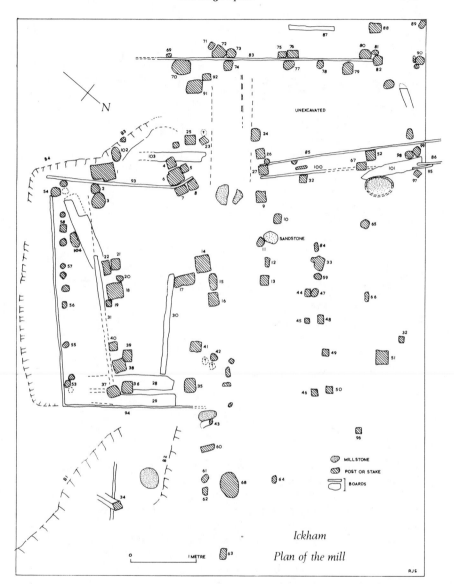

Ickham

Plan of the mill

Figure 2.

estimated although it was probably not much more than 2.0 m, being limited by the proximity of posts 23 and 24 which were most likely, together with post 25, associated with a water control device. Beam 103 might be a horizontal sill for a water-gate or penstock. During the excavation the ground between posts 23 and 24 was found to slope down into the mill and traces of decayed timber, possibly boards, were found lying

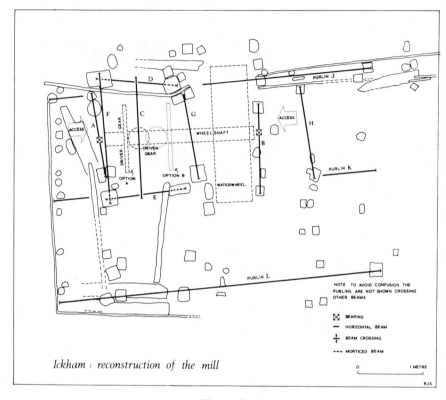

Ickham : reconstruction of the mill

Figure 3.

in this channel about a metre away from the mill wall. This suggests that the headrace was timber-lined close to the mill.

The horizontal shaft on which the waterwheel and driver gear were mounted was supported by two bearings carried by transverse beams, see Fig. 3. The waterwheel was unlikely to be overhung because its weight, especially when running wet, would have probably caused it to over-balance. Moreover, if a bearing had been placed between the wheel and the gear, it would have been difficult to gain access for lubrication and maintenance so it is suggested that the bearings were placed at each end of the wheelshaft. With this arrangement there were two clear advantages: (i) access to the bearings was relatively easy (see Fig. 3, the large arrows) and (ii) the effect on the wheelshaft of differential movement of the transverse beams was minimized.

The axes of the two transverse beams, A and B, are shown but their lengths at this stage of the analysis are indeterminate. The wheelshaft was made of wood[20] and probably had journals of wrought iron inserted in each end. They would be held fast with keep bolts or winged gudgeons and strengthened with iron hoops (*lamminae*) on the shaft ends. Alternatively the journals may have been made of bronze or copper, which was sometimes used for the short shafts of drainage wheels,[21] or less likely of hard wood. The bearing blocks were probably made of hard wood (*stipites*) and lubricated with animal fats or vegetable oils.

For the benefit of this analysis we may assume that an undershot waterwheel with radial floats (*pinnae*) would rotate, under loaded conditions, at approximately one half the average velocity of the water striking the floats.[22] In the absence of figures for the headrace gradient and section, the water velocity cannot be calculated but it is very likely that the millstone ran at a greater speed than a rotary quern. In order for that to have occurred it is probable that the gear ratio was of the order of between four and six to one. Precision on this point is unnecessary; it is merely important that the ratio was significantly greater than unity. Construction details are unimportant to us at the moment as well, the essential point being that the driven gear was considerably smaller than the driver.

To allow the driven gear to engage the driver gear, the vertical spindle on which it was mounted had to be supported by a footstep bearing positioned more or less above the axis of the wheelshaft. This bearing had to be supported by a transverse beam, C, for a longitudinal beam would be obstructed by the driver gear. In modern mills this support beam is called a bridge-tree. The position of the driven gear is an important part of this analysis because it is coincident, in plan, with the spindle and the centre of the millstones above. In order to decide on which face of the driver gear the driven gear probably worked, some regard should be given to the platform layout and meal collection.

As already stated, the millstones were supported on the floor, or platform, above the wheelshaft and gears. The miller required space above the stones for the grain hopper and feeding device and clearance around the stones from posts and the external walls of the mill. When the stones were working, warm meal was being ejected from their peripheries or skirts and this had to be collected regularly by the miller otherwise it would build up and tend to clog the stones. However, it is possible that the miller had an alternative arrangement for collecting the meal. By enclosing the stones with a wooden skirt, called a tun, the meal would be automatically contained and if a hole was made in the floor between the bedstone and the skirt, the meal would readily fall through into a suitable container below. The miller might still be required to clear the meal by periodically pushing it down the hole, and to minimize this he may have elongated the hole. But it would only have been a matter of time before he realized that continuous supervision could be avoided altogether by a simple device. If the upper stone had a projection of wood or iron fixed to its skirt and arranged so that it would pass round the annular space just above the floor, then the meal would be automatically swept through the hole. This device is found in modern watermills and also in the more primitive horizontally-wheeled mills called Greek or Norse mills.[23] At Ickham the meal may have been collected in a bin or it may have passed straight into the sack. A bin, being more open than a sack, would have allowed the meal to cool quicker, but this is not a sufficient advantage to be conclusive.

There seem to be two clear alternatives regarding the position of the driver gear (see Fig. 3). In option (a) the driver gear is positioned as far away from the waterwheel as is practical as determined by satisfactory clearance from post 21. In this case the spindle and driven gear would be on the waterwheel side of the driver gear because we can be fairly certain that the millstones would be contained, in plan, between the four main clusters of posts. Option (b) has the driver gear nearer the waterwheel where its position is limited by satisfactory clearance from post 7 and the wall of the mill. With this arrangement the spindle and driven gear would be away from the waterwheel.

Comparing the two alternatives, (a) is a better arrangement because in position (b) it would be difficult to work on the back of the driver gear and almost impossible to approach the waterwheel from that side. We must remember that access to the side of the waterwheel was necessary to attend to its sole boards, arms and the shaft inside. On the north-west face of the waterwheel, access to the bearing would have been difficult because the bottom half was masked by posts 9–13, 27 and 26. At least one of these posts rose above the level of the shaft and therefore it was very desirable to have access to the waterwheel from the south-east side. To approach the wheel from outside the mill would have meant squeezing between posts 5 and 25 (a 265 mm gap) and then, worse still, between the wheel and post 8 (a 240 mm gap). The wheel may have been slightly narrower but either way access from this side could not have been easy because there must also have been a sharp drop in ground level in the vicinity of posts 5 and 25. A better approach was from inside the mill entering the gear chamber[24] between posts 18 and 17 but this pre-supposes that the driver gear is not in the way. This is why the most likely position for the gear was position (a). In this position both faces of the driver were relatively accessible as was the driven gear and footstep bearing. With the spindle position now determined we may conclude that the bridge-tree was supported by longitudinal beams D and E. To gain access to the gears and waterwheel it would have been necessary to pass under or over beam E which was approximately a metre above the floor of the chamber.

Having identified the primary supporting beams for the machinery we can now turn our attention to the hursting; that is, the frame which supports the millstones,[25] the hopper and the grain-feeding mechanism. The hopper could have been suspended from the roof, for it is unlikely to have been large enough to have the storage function of a true self-clearing grain bin. In addition, the frame would have carried a floor, on which the lower millstone sat, and which also had to take the weight of a man and a sack of grain. In watermills the hursting usually comprises a substantial timber frame, normally a box frame with horizontal beams morticed into heavy corner posts. The millstones would sit on top of the frame and the bridge-tree would be supported at an intermediate level by two side beams. One other feature of the modern mill is worth noting. The bridge-tree normally has a fixed pivot at one end, and is adjustable vertically at the other by levers (to give a mechanical advantage), in order to control the gap between the stones. Such a method of adjustment is unlikely to have existed at Ickham and the footstep bearing was probably elevated by wedges or packing pieces under the bridge-tree.

There are only two alternative methods of supporting the floor over the machinery chamber, by means of either transverse or longitudinal beams laid in parallel. In either arrangement, beams D and E would have to be morticed into posts 1 and 18 and, as there appears to be no disadvantage in having their other ends similarly supported, it is suggested that they terminated in posts 6 and 17. If longitudinal beams were employed there would have been the following disadvantages:

(i) the south-west beam would have been external to the mill wall;

(ii) the north-east beam's position, in the cluster 14–17, would have presented problems. If it rested on post 14, then either the platform had an odd-shaped floor or its boards were overhung. Alternatively, it might have extended beyond post 17 (otherwise the floor would have left a large gap beside the wheel) and if it rested on posts 15 and 16 the function of post 14 becomes problematical;

(iii) the north-east beam would reduce the headroom over beam E where access to the chamber was required.

And so it is probable that the platform was supported by transverse beams F and G carried respectively by posts 1, 18 and 7, 8 and 14. One further advantage can be seen with this layout: transverse beams could be used to lift the wheelshaft for maintenance or replacement of components. For instance, the end of the wheelshaft could be readily lifted from beam F to facilitate replacement of the bearing block or journal.

Although we have identified a symmetry in the post grouping, the inaccuracies of the plan are obvious. It should be remembered, however, that accuracy of plan or construction was not essential to this structure; functional considerations outweighed aesthetics. Space frames do not need to be set precisely because the final position of the machinery is not dependent on the planes being perfectly at right angles to each other. I.A. Richmond has noted another reason for the inaccurate setting of posts in Roman timber structures and that is the tendency for the post to be inserted hard against one side of the pit dug for its installation.[26]

It is interesting to note the large cross-section of the main posts in the hursting and machinery support frame. Typical sections occur in posts 6, 17, 14 and 18 which are approximately $22\,cm \times 24\,cm$ or $528\,cm^2$. Post No. 1 is particularly massive, having a cross section of $780\,cm^2$.

Having established a provisional structural arrangement for supporting the machinery and stones, let us examine a structural problem which was peculiar to the machinery itself and would manifest itself in the area we are currently examining. Each time a radial float on the waterwheel met the water a shock would be generated and a similar effect would originate from the gears as the imperfect cogs engaged. With rotating masses the centrifugal forces tend to amplify any imbalance causing the pressure on the bearing and gears to vary during each revolution. These vibrations and varying stresses would be communicated to the building frame via the bearings and, to a lesser extent, via the platform floor. This incessant pounding of the earth-fast posts over a period of time would be rather like the action of a piledriver or a vibrating road roller and structural movement, mainly in the form of sinking, would result.

It would not have been necessarily harmful if the timber structure as a whole sank a little but what we can be fairly certain of is that different parts of the structure settled at different rates. This differential movement would be caused by the interaction of varying loads and vibrations on various sized posts in different ground conditions. Differential sinking of the structure would be most likely to manifest itself in the mill machinery, as misalignment of the bearings (including the footstep bearing under the spindle), as difficulty in maintaining proper engagement of the gears, or through the millstones requiring repeated adjustment to prevent uneven wear or improper grinding. When this occurred the bearings and the stones were probably adjusted with wooden wedges and packing. If the movement proved embarrassing, the miller or builder might attempt to arrest any leaning of the posts with additional piling, but if the misalignment became very bad the attempt may have been made to introduce new main posts to take the loads. This would not have been easy inside the mill because of the limited space for digging the foundation holes, and if successful it probably necessitated altering or replacing the horizontal beams above. It would, of course, have been much easier to install new main posts outside the mill and this might explain why the large posts 1 and 6 appear to be external.

Structural Analysis—the Floors and Walls

No evidence of clay-daubed or plastered panels of wattling came to light during the excavation so that from the existence of boards 93, 94, 100, etc., we may conclude that this timber-framed building was weather-boarded.

With a waterwheel some 2 m in diameter, the top of the wheel could not have been very far above the millstone platform. If we allow, for the moment, that the waterwheel projected below the ground level of the chamber by 45 cm; that the chamber required a minimum headroom of 1·2 m;[27] that a further 20 cm was needed for the platform and its supporting beams; then the top of the wheel would have been only 15 cm above the platform surface. Such exactness is debatable but it does support the notion that the wheel was floored over by the platform. The miller could not have tolerated an open wheel working close to the millstones because of danger and the admittance of moisture and damp air. He would either have closed it off from the mill with vertical boarding or, more likely, floored it over so that his platform had one constant level. It would also have improved his chamber headroom, given better access to the side of the wheel and given him useful floor space in what was a rather small mill. This floor could be readily supported by beam G and an additional transverse beam H carried by posts 32 and 33.

There is an undeniable temptation to identify boards 28 and 29 and beams 30 and 31 as part of a staircase or steps. In the excavation the boards gave the impression of being undisturbed from their original positions and board 28 was some 75 mm below the level of board 29. If steps did exist here the lower tread boards would be the most likely survivors but no evidence of others came to light. Moreover, the general level of the ground in front of the two boards did not suggest the incline necessary for a flight of steps.

An alternative function for the two boards is that they might be associated with the threshold of a door. Their position, close to the external wall of the mill, is compatible with the suggestion and it was noted during the excavation that the ground outside the mill, on the brickearth bank, was very firm. This was in contrast to the northern half of the site which was much damper and less stable. The obvious entrance, according to ground conditions, was from the south or south-east corner, but definitive evidence for the entrance is not forthcoming; no sill beam was found nor any jamb posts. Posts 35 and 36 might be taken for door jambs but they are displaced longitudinally from the supposed threshold board 29 and, perhaps more telling, board 30 would bar the entrance anyway. Nevertheless, posts 35 and 36 must surely be part of the building structure although curiously isolated from what must be the weatherboards on the east face.

The most significant feature of these boards is that during the excavation they were considered to project beyond post 37 and moreover board 28 was cut to clear post 36. As boards 28 and 29 appear to be associated neither with a door sill nor a flight of steps, we must conclude that they were floorboards *in situ*. Board 31 is clearly an earthboard for taking the pressure of the higher ground against the south-east face. The section through the mill confirms this and boards 28 and 29 pass under 31. When excavated it was found that the south-east wall boards were leaning in, due to soil pressure, and it should be noted that a substantial number of posts occur along this wall indicating that a pressure problem existed here. Observation reveals that posts 53–58 are all spaced the same distance apart, perhaps indicating that these were original and the others subsequent. A more convincing suggestion is that the wall pressure problem was finally overcome by creating a second low internal wall, of which board 31 is a remnant, and by backfilling between this and the external wall. This inner wall or revetment was placed against the main posts 3 and 22 and additional posts 19, 40 and 36 (?) were dug in to give further support. This backfilling would explain why the south-east ends of boards 28 and 29 appear to pass underground.

Board 30 was not an earthboard because supporting posts and a higher ground level are absent and it is unlikely to have been a low safety rail to keep people from falling into the tailrace. Neither is it the bottom board of a wall for it is unusually thick for such a purpose, supporting posts are absent and its position cannot be justified. It may be displaced at its north-eastern end from its original position—perhaps attached to post 35 or 41—but this would break the symmetry of the plan because the transverse axes of boards 30 and 31 are parallel to the south-east wall of the building. Is this fortuitous? The suggestion that board 30 might have been a brayer[28] cannot be sustained. It is far too low for that function even with a vertical push beam at its pivoting end and it takes up far too much space.

Although we are unable to assign a structural purpose to board 30 it is possible that it had a more practical function in the working of the mill. We have established that the meal was delivered into the machinery chamber at a high level and collected either in a wooden bin or in a sack suspended clear of the driver gear in the area between posts 17 and 18. A short chute may have been employed to allow the sack to be positioned away from the gears. If sacks were employed for collecting the meal, board 30 could have been used as a rubbing strake or support-board to give some support to the sack while it was filling and to support full sacks when they were moved away from the meal spout. If this were the case at least one other strake, some 75 cm above the ground, would have existed. Rubbing strakes also have the advantage of keeping the full sacks clear of damp and rough surfaces and this is their main function in modern mills.

At this mill the heavy section of board 30, together with one or more similar weight boards above, would have supported a full meal sack with ease.

The frame of this building was entirely earth-bound; no sole-plates were apparently used in its construction.

The flints that were found on this site do not occur naturally in the local soil so they must have been introduced either for foundation purposes or to create more stable ground conditions. It is not possible to determine with any reasonable accuracy where they were placed, but their distribution is probably indicative of the wet ground conditions that existed on this bank. Of the hundreds of flints found only one rested inside the mill, south-east of the original watercourse, apart from four or five on the very edge of the bank.

There is little doubt that the tailrace would have been boarded over to provide additional floor space; to have left it open in the middle of a working floor was inviting trouble and in winter it would have made the air condition within the mill slightly worse. The most obvious support for the tailrace floor would be two transverse beams, one supported by posts 16 and 35, and the other by posts on the opposite bank. This floor was probably carried across at a level compatible with posts on either side of the water.

During the excavation, post 34 was found with a board attached which suggests that a revetted ditch or watercourse existed there. Two roughly parallel clay banks, B1 and B2, were detected in this region, one on each

side of post 34. This may be evidence of a watercourse or a ditch pre-
dating the mill for it seems to be too far towards the south-east to be part
of the tailrace as defined by posts and stakes 43, 60–63. Post 68 is surely
not a post but a tree that has grown in this position.

One other problem the miller would have had to contend with was
erosion of the water channel within the mill. The consequential question
is, did the waterwheel have a trough[29] or side boards to constrain the
water close to the floats? No evidence of a trough under the wheel or
boards beside the wheel came to light and while posts 26, 27, 9 and 11–
13 could offer the necessary support, their counterparts on the opposite
bank are absent. Of course, there was a difference in ground level either
side of posts 9–13 because the bed of the water channel would have been
lower than the ground each side. If boards existed, posts 11 and 12
probably had the dual function of providing support for waterboards and
revetment for the change in ground level. The ideal arrangement would
have been earthboards on the land side and facing boards on the water
side to promote laminar flow conditions. Beam B was probably carried by
posts 9 and 13. If boards were absent it would have been only a matter
of time before the water scoured the earth from around the posts to
weaken the structure. Without their protection the profile of the stream
would have gradually deteriorated and decreased the efficiency of the
wheel. When this occurred we may credit the miller with enough sense to
have installed at least some side boards in each bank of the channel to
create a crude but effective trough. However, the evidence has been
robbed or, more likely, has eroded away.

At this stage of the analysis an examination of the various levels that
existed within the building may prove helpful. The lowest level, that of
the water channel, is well defined where the waterwheel stood, being
contained within banks on the lines joining posts 23 to 16 and 24 to 13.
But the banks of the tailrace are not so easily defined. Staying for the
moment within the building, the south-east bank surely followed the line
from post 16 to stakes 42 and 43 which were probably associated with
revetment work. The opposite bank appears to be defined by posts 44–50.
Six of these are paired, suggesting either that some of the posts were
introduced subsequently or that this structure withstood pressure from
both sides.

The tailrace would have been at a lower level than the bed below the
waterwheel to ensure that the tail water cleared the wheel. Moreover,
there was probably a gradient for some distance below the wheel so that
in times of spate there was no danger of the wheel working in backwater.

Where water passed through the mill it was obviously constrained by
the banks of the channel and an attempt would have been made to keep
it from splashing into the machinery chamber. It has been suggested that
the ground level of the machinery chamber was some 45 cm above the
bottom of the watercourse alongside and that the wheel may have operated between side boards, which for water-constraining purposes would
not have needed to be much above ground level. However, while the

miller would have needed access to the wheel for inspection and maintenance he could have installed low boarding beside the wheel to stop water splashing into the chamber, for his sacks were being filled only some one and a half metres away. The ground of the chamber, which may not have been boarded, appeared to be lower, perhaps as much as 20 cm, than the floor represented by boards 28 and 29. It may have been designed this way so that any water gaining entry from the watercourse stayed within the chamber and drained back into the tailrace.

The general floor level, as far as we can tell, was boarded between the area bounded by posts 20, 36, 47 and 50, and continued at the same level with flints out to the south-west and north-west walls of the mill.

The next level within the mill was the relatively long narrow backfilled strip bounded by posts 2, 37, 53 and 54. It is not easy to suggest how high this stood above the main floor because of the erosion that has taken place, indeed it may have varied or sloped, but it is considered to have been at least 300 mm high. It was from this level that the wheelshaft bearing, at the south-east end of the shaft, was maintained.

The fourth and highest level that we are able to identify is the millstone floor or platform which we have concluded was extended over the wheel and was bounded by posts 1, 18, 32 and 33.

The Roof

Roof tiles were found widely scattered all over this site but there were insufficient numbers close to the structure to support the suggestion that they originated from the roof of the mill. Their wide distribution has prompted the idea that they originated from some other building nearby, yet to be discovered or perhaps already lost. In support of this notion is the point that if tiles were used it is difficult to see why some tiles did not find their way into the south-east half of the mill, and so it is thought that the mill was roofed by thatch and not tiles.

In several ways, thatch had advantages over tiles. Even allowing for the greater thickness of thatch, whether reed or straw, its density, of the order of 1/7th that of burnt clay, resulted in a lighter load on the structure. In view of the wet nature of the ground the builder would obviously have wanted to minimize the weight of the roof. Moreover, thatch was able to absorb structural movement, whether vibration or settlement, far better than clay tiles. It could also take a higher pitch angle. And in any case, there is no reason to suppose that our builder looked any further than the natural resources of the area, although he may, out of habit of reaping high with a sickle, have used tall stubble. Of the straws, rye is the longest and strongest while heather does not rot so quickly. Sedge has the advantage, with its sharp serrated edges, of keeping birds and vermin away but the material that the builder most likely used was the reed which probably dominated the area. The thatch was most likely carried on laths but it was not unknown for wattlework[30] or brushwood to be used.

If the proposed layout of the levels within the mill is correct, the most practical arrangement for the roof would have been a single pitch, with its high edge on the south-west side where headroom was required above the platform. This would have shed water onto the tailrace side of the building where it would readily run away; on any other face it would have required land gutters to carry it away. No evidence of drip-trenches or timber gutters was found. Rain-water coming off the south-west side of the mill had to enter the head-race which might have been difficult to arrange because of the revetment or side boards that probably existed. With a north-east orientation, a single pitch thatched roof would also have had the advantage of protecting the mill from the winter weather blowing across the Wanstsum marshes (see Fig. 4).

Ickham: an artist's view looking west   RJS 1984

Figure 4.

How then was the roof carried? With transverse rafters (parallel to the stream) their span would require end and middle support so that we should expect posts for supporting three longitudinal purlins. The posts most likely to have had a roof support function are numbers 54, 52, 18, 33, 37 and 51 supporting purlins J, K and L. It is significant that these three purlins are parallel and one *passus*[31] apart from each other. Post 15 may have acted as a mid-support for purlin K and similarly posts 26 and 27 probably rose to meet purlin J. The fact that they are apart may, however, indicate that they straddled the purlin. Post 35 is unlikely to have supported purlin L because of its proximity to post 37.

So far this structural analysis has assumed a 'post and truss' principal but we should be prepared to consider that curved principals may have existed. The fact that inclinations have not been recorded on the post fragments is not indicative of their absence. Cruck-type principals spring from near-vertical bases. Nor can we be influenced by an earth-fast frame which might suggest that braces were largely unnecessary as would the heavy post sections which may have been selected due to the wet nature of the ground. But it is difficult to imagine the complete absence of wind bracing on such an exposed site.

The use of curved principals does not necessarily rule out a single pitched roof, for there may not have been symmetry of elevation. Indeed, it is difficult to identify any symmetry in the plan of the building and hence there is no reason why it should occur in the section.

It is difficult to suggest an entirely satisfactory arrangement for curved principals had they existed, but if used in juxtaposition with vertical posts for horizontal tie beams or possibly wall plates, three areas may be tentatively suggested: in the area of posts 36–39; at 1 and 102; or at 26, 27. But the problems which this hypothesis creates for the roof design are too numerous for a satisfactory analysis. Perhaps a reasonable conclusion would be to accept that it is more difficult to integrate into a building with curved principals the frame and hursting, which are in essence themselves small post-and-truss structures, which support the machinery.

Possible Extensions to the Mill

The axis of posts 47–50 is intriguing; it lies parallel to the transverse axis of the mill. Furthermore, if this line is extended it passes very close to posts 59, 33, 84 and 32. In terms of the constructional accuracy that we are dealing with they may be regarded as lying on one axis. Then there is the question as to why the tailrace appears to widen so quickly—it doubles its width in 30 cm of travel. One possible explanation of this is that the north-west wall of the mill originally stood on the line between posts 32 and 50, and at some later date the mill was extended to posts 51 and 67 or 52. If this were so the original main post in the north corner may be missing; posts 46 and 50 are rather too light in section to have performed that function.

If an extension was made, might not this explain why flints were brought into the mill on the north-west bank? The obvious reason for an extension is the need for greater floor space but it may have been a response to the tailrace eroding its north-west bank. When this mill was originally built, the watercourse would have had a constant width through the mill and yet the flint scatter shows us that the tailrace moved subsequently towards the north. The posts that must have existed between posts 13 and 64 have disappeared. Of course, much of this movement of the tailrace must have occurred after the mill was abandoned, but some erosion probably took place during occupation. For this to have occurred, the tailrace must have been without proper revetment for the miller would not have allowed reveted courses to decay so. In moving, the tailrace may have eroded the bases of the support posts of the original north-west wall, prompting the miller to extend the building and utilize the old mill wall posts for defining the tailrace bank. Posts 44–46 may have been introduced to give additional support when the new floor was made up with flints but more likely they were associated with new revetment work introduced on this bank.

When looking at the roof plan and, in particular, the overhang of the three purlins on the south-east side of the building, one other possibility suggests itself—that the mill may have been extended on its south-east face. If the original wall was on the line of posts 2 and 37, posts 19 and 40 might be vestiges of the original wall even though we have already ascribed satisfactory functions to these in suggesting that they supported earthboard 31. It is interesting to see that bank B1 (see Fig. 2) appears to terminate close to the mill on the axis of posts 2 and 37. Moreover, bank B3 appears to do the same. These might be the original edges of the excavation made for the earlier wall prior to the extension of the south-east face.

When this extension occurred the builder may have thrown the excavated earth towards the south-east, thus inadvertently causing his next problem—increasing ground pressure on the new wall. The events that followed—installation of more wall posts and backfilling with an internal revetment—have already been suggested. Besides an increase in floor area, the south-east extension gave the miller the advantage of being able to reach the south-east wheelshaft bearing from inside the mill. Hitherto access was from outside the mill, a fact which must have inconvenienced him whenever adverse weather prevailed, especially since the other bearing was accessible from inside. Perhaps this is why the depth of the south-east extension is the same as the original access depth on the north-west side; that is, between beams B and H. This depth may have been influenced by the amount of overhang that the builder considered he could tolerate without the need to erect new main posts.

Boards 93 and 94 appear to be in single lengths which means that they must have been installed when the south-east extension was made. The kink in the end of board 93, between posts 6, 7 and 8, defies explanation; presumably the axial misalignment of posts 4–8 affected most of the weatherboards on this face making their replacement very difficult without post removal. To install new ones would have necessitated steaming the ends to shape and inserting them between posts 1 and 2. Perhaps the kink only occurred in the bottom one or two boards close to the ground and the others above were nailed onto the overlap of post 6 beyond post 7.

We do not know if these extensions to the south-east and north-west occurred at the same time but it is clear that each necessitated changing the three purlins for longer ones so that it was advantageous to extend both sides simultaneously. With a lengthening of purlins the entire roof would require removal, including lathes and rafters, and this may have coincided with re-thatching. It is relevant to note that the best thatching, reed, lasts for 50 years, some claim longer, whereas straw lasts from 10 to 30 years. Although much depends on the pitch and aspect of the roof, and of course the workmanship, these figures ought to be considered as an upper limit and the conclusion must be therefore that the thatch of the mill was probably changed several times during the lifetime of the building.

If the south-east extension was done separately, we cannot be sure that an entire roof removal occurred; the builder may have attempted to introduce additional shorter purlins supported by two posts with the new roof overhung. This might explain the function of posts 35 and 16 which could have supported the north-west ends of the additional purlins. The new purlin could have shared post 6 with purlin J. But this arrangement would have been difficult to execute for it would have required partial stripping and/or lifting of the roof to insert the new purlins. Furthermore, to ensure success the new purlins had to be tied to the old and anchored either to the new support posts at their north-west ends or to the overhanging purlins supported from the new south-east wall. The absence of heavy sectioned posts in the wall suggests that it was not intended to take the load of the roof. If neither of these options was adopted, the existing rafters would have been levered up and an undue weight put on the new wall. Given such complications the sensible solution would have been the replacement of the existing purlins and this would most certainly have been the case if the two extensions were executed simultaneously.

With a single pitch roof of north-east orientation, extensions on the south-east and north-west could be undertaken without altering or replacing the main posts. However, with a single pitch roof of south-east or north-west orientation, an extension on both of those faces would have necessitated alteration of the main posts, unless the miller was happy with the very low headroom that would have resulted.

Further Speculations

In the discussion so far, functions have been ascribed to all large and medium section posts within the mill except Nos 10, 104, 38 and 39. It should also be noted that posts 16, 41, 35 and 36 have fairly large sections which perhaps exceed that required for the floor-supporting function that we have given most of them. This prompts further speculation.

Post 35 might, just, be an original roof post if the builder repeated the three bays of the south-west face on the north-east side; that is, the counterpart of post 35 would be the group centred on post 6. But neither the original purlin, nor its replacement L, would require a support at post 35. The heavy section of post 41 has been explained as a support for a floor beam taking the weight of grain or meal sacks. But what of 38 and 39? Should 36 be queried?

It might be assumed that post 36 should be taken with 37 as a roof support, just as 26 and 27 are paired, or 67 and 52 in the south-west corner of the building. But caution should be exercised here; these pairings may belie the obvious. It is interesting that all of these posts have very similar sections, fairly substantial but not quite as large as 51 which certainly is an isolated roof post. Post 67 is surely a wall post for it aligns with 66 and 65, and it is diagonally placed to 52 so that it could not share the support of purlin J. Post 52 might have been a later corner post to replace 67, which would certainly have decayed long before corner post 51. Posts 26 and 27 deserve discussion. Although it is easy to suggest that the south-west wall had three bays with the watercourse passing through the middle one, the span of purlin J would not require support at each bay division. Post 27 must be, by virtue of its position next to the water-wheel, a wall-end post. But post 26 has a more interesting function when we consider it as a counterpart to posts 23 and 25. It is very likely part of a frame for either a water-control gate or a water-trough. So having rejected the idea of paired roof posts, post 36 can now be included in our speculation of that area.

Two solutions may be put forward for consideration. The first, and most probable, is that the platform or millstone floor was extended to posts 35 and 36. Such a floor would be carried on two beams, one resting on posts 16, 41 and 35 and the other on 18, 38/39 and 36. Beam F could have served as one support if it was longer but the same cannot be said for beam G which is misaligned for such a purpose. The second solution, less likely though more exciting, is that these posts may have formed support frames for the machinery and millstone of an earlier waterwheel. One can tentatively identify the support functions of the posts but the problems raised by this arrangement become complex and the solutions more speculative. Although these posts display some symmetry in plan, a symmetry which is a pre-requisite for machinery, it is not possible to define a more precise arrangement because of the subsequent alterations, and therefore a 'coherent' history of the posts in this area is not possible. The suggestion of an earlier waterwheel and shaft position cannot be ruled out, but it is ultimately impossible to justify in relation to the projected history of the mill.

Posts 10 and 104 occupy peculiar positions. Because they lie roughly on the wheelshaft axis and are fairly close to the beams A and B, they might be taken as support posts and yet they are displaced from the axes of the beams and bar the access space to the bearings. Because their position is not coincident with the hursting or building frames it is not possible to assign a function to them related to either the platform or the roof so it is unlikely that they were higher than the bearings. The fact that there is one near each end of the wheelshaft, close to the bearings, must be signi-ficant but we can only speculate about their function. They might have been associated with the lubrication of the bearings but it is difficult to see why the bearing support beams could not have served this purpose themselves. Alternatively, they may have been curved braces for beams A and B to prevent longitudinal movement of the wheelshaft.

When searching for an explanation as to why principal posts 1 and 6 are external to the mill, it seems very likely that the builder may have underestimated the structural load and/or the ground conditions. The key to this design failure is the multiplicity of posts that existed inside the mill at each corner of the hursting, viz. posts 21 and 22; posts 2 and 3 and posts 7 and 8. The requisite number of posts existed to provide adequate support without posts 1 and 6. The movement of the frame would not have shown immediately, but when it became pronounced major structural modification was avoided by digging in new principal posts external to the mill. The substantial scantling of post 1, and the group piling of 4–6, which are greatly in excess of normal structural demands, shows that the problem was severe. If these suggestions are correct, posts 18 and 17 were probably introduced at the same time, and it is possible that post 14 replaced two posts of smaller scantling.

This major upheaval necessitated installing a longer beam at F and a new bridge-tree C; it may have been possible to re-use, with modification, beams G, A, D and E. The insertion of post 18 may have given the opportunity of making the south-east extension, especially in view of the fact that the original roof post 2 could be superseded by post 54 and the builder could cut down post 2 and rest beam B on it to help overcome settlement.

An alternative but less likely explanation for posts 10 and 104 might be that they were datum points against which to estimate settlement.

Having defined the various floor areas and levels that may have existed in this mill, it remains to consider the location of steps or stairs which connected the ground floor and the platform. There are several alternatives, each having inconclusive minor advantages and disadvantages, and unfortunately it is not possible to decide with any confidence where they might have been situated. It would be convenient to propose that posts 38 and 39 were connected with stairs and this might explain their proximity to posts 36 and 37, but really no satisfactory arrangement can be deduced. A similar difficulty is placing the door. As already stated, no sill board or jamb posts have been found or identified and thus it is not possible to determine where the entrance was.

The South-West Structure

Our attention must now turn to what is probably the most difficult part of this analysis, to ascribe a function to the structures lying to the south-west of the mill.

Board 83 lies exactly at right angles to the head-race and is held fast by many medium to large posts. Their positions either side of the board show that they were placed to withstand movement or pressure from both sides. If this was the base of an earthen dam there would be no need for the waterside posts because backthrust would not occur. A pond would have had advantages. The first would be an improved head of water so that delivery was from a higher level. The second advantage would be that a pond gave a reserve for times of reduced flow or non-working periods. In other Roman watermills, when a head of water was created to allow the use of an overshot application, such as at the Athenian Agora Mill,[32] the builders appear to have been primarily interested in creating a high water velocity. No attempt was made to create storage. Even at the extraordinary arrangement at Barbegal, the small area of water where the steam expands to feed the parallel head-races either side of the mill cannot be taken as a pond.[33] At Ickham, the velocity of the water approaching the wheel could be increased either by a pond or an elevated head-race. To create an elevated watercourse in the relatively flat land surrounding the mill would have meant travelling a considerable distance upstream to create a positive head at the mill, while the alternative of creating a dam would have necessitated throwing up banks not only across the stream but along both sides as well. No evidence was found of any upstream structures or earthworks to confirm the existence of a dam. Furthermore, if the south-west structure *was* intended for a dam there would surely be some evidence of it on the alignment of the head-race where the water passed through or over the dam. But there is no break in board 83 nor any posts which align with the sides of the head-race. We can also dismiss any idea that the head-race was deflected just before it reached the wheel; this would have been impracticable running so close to the mill, and loss of velocity and deposition would have occurred.

It might be thought that posts 82, 88, 89 and 90 could have formed the frame for a water control device at the beginning of the bypass through which any water in excess of the mill's requirements was uallowed to flow. And while there must have been such a bypass no trace of its course was found in this area.

And so the suggestion that the south-west structure was part of a dam is negated by the absence of supporting evidence and the presence of posts on the south-west side of board 83. A far more convincing possibility is that this structure is not related to the working of the mill at all. If this were so the structure would not antedate the mill—it violates the head-race course—and so it must have been built after the mill ceased to function.

When comparing the south-west structure with the original mill, an interesting fact emerges. All the posts inside the mill, excluding the stakes, have regular sections of four dressed faces except 47, 2 and 3. It is notable also that even some small posts are squared, for example 43–50, 19 and 40. Examining the south-west structure, however, one difference is very noticeable, namely the posts in contact with board 83 of which seven out of thirteen are irregular in section. Could this be taken as an indication that it was constructed by different builders at a later date? At this juncture a further fact should be pointed out—the south-east wall of the mill, which we have tentatively identified as a subsequent extension of the mill, also has a high proportion of irregular or rounded posts. This tends to confirm the suggestion that unsquared posts are later work.

The south-west structure was certainly influenced by the mill for its axis is parallel and so it must have been built while the mill stood, but what was its purpose? Board 83 is not a weatherboard; an external wall would not require such large posts on both faces. Why has the south-west wall of the mill a cavity where it faces the south-west structure? It is surely

Ickham : plan showing flints and tiles

RJS 1984

Figure 5.

no accident that board 87 and posts 88 and 89 are on an axis parallel to
board 83 and post 90, but what is the significance? Although the conglom-
eration of boards and posts surrounding post 99 are external to the mill,
they were obviously influenced by its position; boards 95, 86 and 101 are
a continuation of the mill wallboard 100. Board 85 was surely installed at
the same time as this group and it is conceivable that they were all
connected with the south-west structure. For that reason the development
of this area seems to be more explicable if these elements, including board

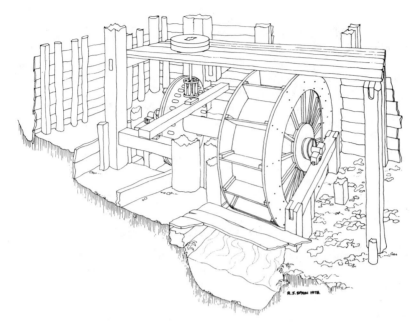

Figure 6. Ickham: artist's view inside the mill.

85, are considered as additions to the existing mill structure. If this was the case, then board 100 was obviously in place, because the cavity created by adding 85 was continued beyond the corner of the mill with boards 95, 86 and 101. This means that the wall above board 100, including posts 27, 32 and 67, was present when these external posts and boards were installed.

If the south-west structure was erected after the existing building had ceased to function as a mill we need not be too concerned about its purpose. The area enclosed by the south-west fabric does not appear to have been related to the mill machinery, nor has it apparently utilized the watercourse for power purposes. Indeed, one is tempted to suggest that the millstream may have been diverted, by man or nature, before the south-west structure was built. But its function will probably remain a mystery. The unexcavated area between boards 83 and posts 77 and 82 may have held further clues to this problem.

When the Ickham mill was abandoned, the stream course shifted slightly, still entering the mill at the same point but veering towards the north in the tailrace. The flints were scattered and scoured away from the final bed of the stream. The axis of erosion and scatter seems to be from the south-west to the north-east though the final axis, marked by the flint-free stream course, was nearly south to north (see Fig. 5). Many flints occur outside of the mill on the south-west side indicating that some drift must have occurred towards that region.

The subsequent erosion drifts of the abandoned site must also be indicated by the distribution of millstone segments. Apart from one or two segments, the greater number lie on an axis south-west to north-east. They extend from the waterwheel position towards post 63. Much of this material probably originated from broken or worn millstones discarded by the miller, quite likely thrown into the tailrace. When the structures on this site were abandoned, much of the timber work may have been robbed for no horizontal beams were found in or close to the mill. The few boards found also indicate that much of the flooring was probably removed or swept away in times of flood.

Figure 6 is a conjectural view of the inside of the mill looking towards the machinery and waterwheel. The Appendix to this study contains notes explaining the design of the waterwheel and gears.

Summary of the Building History

The above analysis and discussion concerning the development of the site may be summarized in the following divisions or phases which are presented in the historical order conducive to the facts (see Fig. 7).

PHASE ONE

The plan of the original mill may be identified by corner posts 2, 32 and either post 36 or 37. The north corner post might be 50, but it is possible that the original is absent from the plan. The mill stream would have passed through the mill on a more or less straight line having a constant width. Its south-east bank followed the posts numbered 92, 23, 15, 42 and 62 and the north-west bank posts 24, 9, 13 and 64.

PHASE TWO

The mill was extended on its north-west side out to new corner posts 67 and 51 and a new wall built including studs 65 and 66. This extension would have necessitated new longer purlins and they probably coincided with a new thatch. The excavation indicates that the north-west bank of the tailrace suffered from erosion. Some of this undoubtedly occurred after the mill was abandoned but it is possible that the erosion began during occupation, probably weakening the bank and the posts of the north-west wall. This may have prompted the north-west extension. Parts of the old north-west wall frame were retained for stream revetment work and the floor within the mill, hitherto boarded, was supplemented by flints in the extension.

PHASE THREE

Probably not more than a few decades after the mill was built,[34] the structure supporting the machinery and stones decayed, settled and moved to such an extent as to affect the grinding operations. The miller

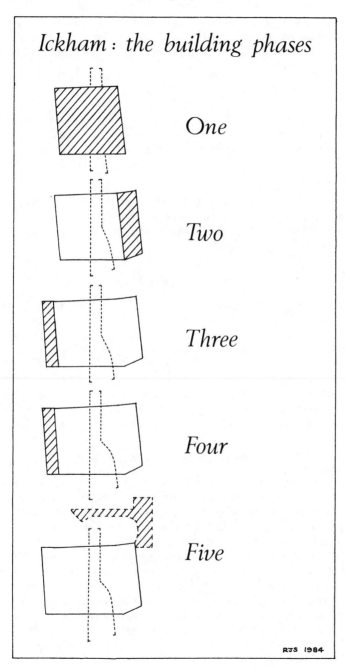

Figure 7.

introduced new external support posts for the roof and machinery, and modified the hurst to suit. At about this time he extended the mill on the south-east side, installing new roof purlins; again this probably coincided with a new thatch.

PHASE FOUR

The new south-east wall of the mill suffered from excessive ground pressure which the miller attempted to overcome by creating an internal revetment on the original south-east wall line and backfilling inside.

PHASE FIVE

It is suggested that phase five, the south-west structure and boards and posts surrounding post 99, were built after the mill ceased to work. Their purpose is unknown. There are indications that the workmanship involved was inferior to that of the mill.

Dating of the Mill

Evidence for dating the mill comes from the 93 coins found within or close to the structure. These have been identified, classified and compared with histograms both of the entire site and of Arles, in the south of France, as a datum.[35]

The conclusion from this investigation is that the mill was probably built and in operation close to AD 150 and following a period of approximately 130 years' usage had ceased working by AD 280. Although fourteen fourth-century coins were found in this area, their distribution on the histogram suggests that they were not the result of sustained occupation.

The Fourth-Century Watermill

It is not the intention to examine the later watermill in any detail within this paper. The analysis of the structure and finds is still in progress and will undoubtedly be the subject of a separate report. However, it will be helpful to this study if some mention is made of the later mill.

Some fifty years after the earlier watermill was abandoned, a larger mill was built approximately 140 m upstream from the old site. The structure of this timber-framed mill appears to have been generally earth-fast and, as in the earlier mill, numerous well-preserved timbers were found in the waterlogged ground. A most interesting and dominant feature of the excavation was the revetted millstream which was floored with planks fixed to uprights and stretchers to make a race about 1·4 m wide and at least 28 m long. Subject to further analysis the mill building appears to have been some 13 m in length beside the stream. We must therefore consider that there might have existed more than one waterwheel. A concentration of millstone fragments were found within the channel confirming that corn-milling was carried on.

A second water channel was discovered some 20 m to the south of the main one, revetted with hurdles and fastened with stakes in the banks. Within this channel the remains of a substantial timber sluice-gate were found which suggests that this watercourse was the bypass for the mill.

The area adjacent to the mill produced considerable evidence of industrial activity. Great quantities of metal and a large number of fourth-century coins were found in the channel and numerous artefacts made of iron, pewter and copper alloys came to light. Among the finds were three bearing stones for journals and an iron hammer-head exhibiting patterns of mechanical wear which prompts the suggestion that water power may have been applied here to metal working. Several brooches, belt buckles and fittings that were found close to the mill together with some lead seals found nearby have been related to the possible existence of a military or civilian official authority on this site. Young has suggested that possibly a works depot existed at Ickham, to service the Saxon shore forts of East Kent.[36]

The Millstones and Querns

The two mill sites yielded numerous fragments of millstones and querns together with two hammer stones. An analysis of the stones[37] revealed that many had suffered degradation of their original faces and profiles by subsequent change of use. In determining the original use of the fragments, twenty-three different millstones, twenty-eight rotary querns and six saddle querns have been identified. Many of the rotary querns exhibited degradation to saddle work and often, combined with further fragmentation, whetting. Twenty-one other stones were found where the original function could not be determined but many showed evidence of grinding and/or whetting.

Six of the larger millstones, all from the later watermill, exhibited a sufficient area of dressed furrows to determine the stones' direction of rotation. Some were dressed for clockwise operation and others for the opposite direction and, although alteration to the gearing arrangement during the life of the mill could explain this, it could also be taken as evidence that two or more waterwheels were in simultaneous operation.

Only two stones exhibited evidence of emplacements for mill-rynds, both two-winged. Two-thirds of the millstones were made of greensand, mainly the lower variety, while the majority of the rotary querns, four-fifths of the saddle querns and five-sevenths of all the whetstone work was associated with millstone grit. Among the other types of stone was a single specimen of Mayen lava.

Conclusions

The analysis of the second-century mill at Ickham has shown that the structure comprised a wheelshaft supported on parallel transverse beams with the vertical spindle supported by a transverse bridge carried on parallel longitudinal beams. The four horizontal beams were supported by separate posts at each end, making a total of eight posts. The millstone platform was carried above the wheelshaft and bridge-tree by transverse beams, one carried by two of the bridge posts, and others carried by separate posts. In modern mills, all four of the bridge-tree posts rise to support the stones and platform, and the beams supporting the bridge are more elaborate with sliding tenons, brayer and steelyard for vertical adjustment of the millstone spindle. The Ickham frame then is not identical to the traditional hursting because the support function of the bridge-tree and millstone platform are not completely integrated.

With the waterwheel external to the hursting, the wheelshaft support posts stand on opposite banks and one pair is remote from the frame. But where the power shaft does not require a bearing external to the hursting, complete support integration of platform, bridge-tree and wheelshaft is possible. This is the conjectured arrangement for the Zugmantel spindles which, due to their origin in a Roman fort high in the Taunus mountains, are believed to have been man-powered.[38] Of course, without an overhung waterwheel it is impossible to combine all three primary support functions unless the frame is unusually large.

In the Zugmantel reconstruction by Jacobi,[39] the wheelshaft is replaced by a horizontal shaft, with driver gear, supported by cross beams between the corners posts and driven by a long-handled crank external to the frame. As at Ickham the bridge-tree is supported by parallel, longitudinal beams between the corner posts. Although the model arrangement is entirely conjectural, though probable, the evidence at Ickham is more tangible—the position and sizes of the posts are known.

For the Athenian Agora watermill,[40] which was in operation much later than the early Ickham mill,[41] Parsons suggested that the wooden wheelshaft journals[42] rested on wooden blocks held in stone sockets and implied that the millstone spindle was carried on a square frame surrounding the shaft.[43] His reconstruction of the millstone platform allowed the timber frame to rest on a sill beam on one side and beam holes in the fabric elsewhere, an arrangement not without its problems.[44]

More recently, Schiøler and Wikander have produced a most interesting reconstruction of the watermill that once existed within the subterranean chambers beneath the Baths of Caracalla in Rome.[45] Substantial brick-faced concrete walls and plinths remain which show that the wooden wheelshafts[46] and bridge-trees were supported directly from the masonry. Stone sockets for the wheelshaft bearings exist, similar to those found at the Agora mill, as do the emplacements for the ends of the bridge-trees, hewn in travertine blocks. The level of the millstone platform has been determined by careful observation of lime deposits and it would appear that its supporting joists probably rested directly onto the adjacent concrete walls so that hursting was absent.[47]

The life of the Ickham mill was largely determined by the rate of decay of its earth-fast timber frame. Modifications to the structure to arrest the settlement probably tended to lengthen the life of the mill as did the extensions, but there were still many original posts which must have reached the end of their useful life long before the introduction of their replacements. Another determinant of the mill's life was the rate at which the watercourses decayed and became less efficient as erosion and deposition took place. In terms of river régime, this stream was relatively young, as was the Stour, and it carried great quantities of silt in suspension. Deposition occurred wherever the velocity was reduced, and as the mill was near the mouth of the stream, silting-up of the courses must have been a continual problem. It would have been particularly bad under the wheel and in the tailrace. Unless the miller frequently dug out well downstream, the deposits would have built up and become overgrown to an extent that must have caused problems. In such circumstances and faced with widespread structural decay in the mill, the obvious solution was to abandon the site and build a new mill upstream there incorporating improvements in its design.

A suggestion has been made that the later Ickham mill may have served the Roman stations at Reculver and Richborough. The scale of industrial operations and the official nature of some of the artefacts do not apply to the earlier watermill which was much smaller in size and, subject to further analysis, its output of meal was probably much less. Moreover, the mill was too small to have stored indoors more than, say, twenty sacks of grain even after the extensions were made.

In attempting to determine what role it may have played in the economy of the area, consideration must be given to the mill's geographic situation. When considering the ergonomics of corn-milling, we should remember that grain can be readily stored for lengthy periods whereas meal and flour spoils quickly. Watermills would therefore tend to be positioned nearer to the product consumers than the corn producers. Accepting that the finding of suitable mill sites was not a problem in Britannia, the choice of mill position was largely influenced by the nature of the market and transport considerations. The presence of other watermills, which could well have existed and operated in the area, might also have affected the transport arrangements.

It is interesting to note how close this mill was to navigable water (see Fig. 8). As already stated, the present tidal head is only one kilometre away from the mill and in Roman times it must have been much closer, probably above the ford. Whether or not the river was navigable above the ford is unknown, although some indication might be given by the fact that the Roman quays serving Canterbury were at Fordwich where presumably the width of the Stour was much greater than the branch of the Little Stour with which we are concerned. If a quay existed it probably lay just north of the ford on the west bank and this would explain why the mill was sited on the unstable marshy ground and as far east as possible in order to reduce the distance to the quay.

Finally, we must ask ourselves who built this mill? The plan and building of the structure do not appear to have had the accuracy and workmanship that we tend to associate with the Roman military mind. It was

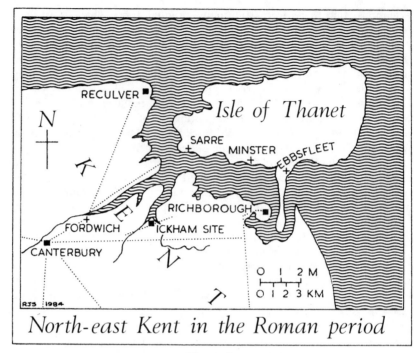

Figure 8.

ill-sited and the builders were not aware of problems peculiar to water-mills. The foundations were laid out rather inaccurately and the posts associated with the machinery were unsymmetrical, surprisingly so in view of the fact that they were dug in. The dressing of many of the posts was crude and the selection of their scantling in relation to their function somewhat haphazard. One has the feeling that there was little planning and that the builder was working intuitively rather than from a prepared plan. All this suggests that the military or Roman involvement was some-what less than the native participation and the most likely explanation is that the mill was executed largely by native labour, perhaps under Roman supervision, to serve the demands for meal and flour of nearby forts and villas. The consensus of evidence suggests that it was probably not an official establishment.

The evidence from the earlier Ickham mill site is unique; it is the first wholly timber-framed watermill found in the Roman world and one of the oldest. It is the only Romano-British example which lends itself to a detailed reconstruction. Only three other watermills are known in the Empire for which there is sufficient evidence to allow a reconstruction of the machinery,[48] and it is primarily for this reason that the early Ickham mill should be viewed as an important contribution to our knowledge of Roman watermills.

Acknowledgements

I am indebted to Mr J. Bradshaw, director of this excavation, without whose help this study could not have been made. As the author was not present during the excavation, Mr Bradshaw's first-hand knowledge was, to say the least, invaluable. During numerous and often exacting enquiries he showed great patience and always transmitted his knowledge in a most lucid manner, for which I am deeply grateful.

My sincere thanks to Dr P.H. Draper and Mr D. Kelly for their reading of my manuscript and giving me the benefit of their constructive criticisms, and to Mr K.G. Elks for his interpretation and dating of the coin evidence.

Finally I must thank Dr C. Young of the Department of the Environment for allowing me to use drawings and other information from the excavation, and Dr. N.A.F. Smith of Imperial College, my research supervisor, for his guidance on the structure of this paper.

Appendix

It has been decided to show the waterwheel construction (Fig. 6) in some detail in order to add realism to the view. This has the disadvantage of being more speculative than the remainder of the illustrations and some explanation of the arrangement is considered necessary.

No remains of Roman waterwheels have ever been found although an impression of a small shrouded overshot wheel was found at Venafro[49] and the diameter and width of several have been determined at other sites.[50] However, the remains of many man-powered wheels used for draining Roman mines have been found and the designs of these have influenced the author's reconstruction. In particular, the drainage wheels from Tharsis,[51] Rio Tinto in Lusitania,[52] and Ruda and Verespatak in Dacia[53] are considered to be of especial interest in this respect. But it is important to note that although such wheels reflect Roman ingenuity and technique, they are the product of a different environment compared with water-powered wheels. Drainage wheels tended to be very economic in material because they were intended to elevate as much water as possible for a minimum of effort. This has resulted in very economic designs. Furthermore, as they were invariably made before being taken into the mines where they were to be assembled, accuracy of manufacture, stemming from the necessity of pre-fabricating parts, was greater than would normally occur. The design of drainage wheels was therefore probably in advance of water-powered wheels.

An early fifth-century mosaic of an undershot wheel found at the Great Palace of Byzantium,[54] although somewhat later than the period we are concerned with here, shows radial arms and a generally lighter frame rather than the heavier clasp-armed frame often associated with Saxon waterwheels. Radial arms also occur in the painting of an overshot waterwheel in the catacombs of Santa Agnese in Rome which Schiøler claims for the third century AD.[55]

One other major feature which has to be decided is whether the wheel had open floats or was enclosed on each side by shrouds. The author favours a waterwheel having shrouds the full depth of the floats for the following reasons:

(i) an annular rim of some sort would have been needed to stabilize the arms and floats;

(ii) although the above reconstruction concluded that side-boards probably existed on the banks either side of the wheel, it is thought that these were installed to reduce scouring rather than to create a close-fitting trough.

We must next ask ourselves if the wheel had sole-boards, that is boards between the shrouds at their inner diameters and at the ends of the floats nearest the wheelshaft. When the depth of water entering the wheel was near or equal to the float depth, the water would have tended, in the absence of sole-boards, to rise over the floats. The addition of sole-boards would have stopped this and reduced the amount of water splashing into the mill. The main disadvantage of sole-boards, aside from the slight increase in wheel weight, was that if the millstream was in spate there was a danger, unless the flow was effectively controlled, that the surge of water would lift the wheel from its bearings. This is more likely to occur where an undershot waterwheel works within a large stream or river and, on balance, it is considered that the Ickham wheel probably had sole-boards.

Bearing in mind the quality of the workmanship displayed in the building, the author considers that the driver gear should take the form of the cruder pin-type gear rather than a rectangular cog-type. The lantern-shaped driven gear is based on the Zugmantel specimen.[56]

Notes and References

1. F.G. Simpson, *Watermills and Military Works on Hadrian's Wall: Excavations in Northumberland, 1907–13*, edited by Grace Simpson with a contribution on watermills by Lord Wilson of High Wray, Kendal, 1976, pp. 26–42.

2. R.C. Shaw, 'Excavations at Willowford' in *Transactions Cumberland and Westmorland Antiquarian and Archaeological Society*, Vol. xxvi, 1926, pp. 429–506.

3. J. Clayton, 'The Roman Bridge of Cilurnum' in *Archaeologia Aeliana*, Vol. vi, 1861, pp. 80–5.

4. I am indebted to Mrs H. Dean Hughes (née Pettit) for bringing to my notice the information concerning this mill site. A report of the excavation work and finds does not appear to have been published.

5. Fullerton Villa, Hampshire, *Journal of Roman Studies*, vol. lv, 1965, p. 217. The excavation was directed for the Ministry of Public Building and Works by Mr and Mrs D.B. Whitehouse. Information was provided by Dr Whitehouse in private correspondence with the author.

6. W.J. Wedlake, *The Excavation of the Shrine of Apollo at Nettleton, Wiltshire,*

1956–71, Society of Antiquaries, London, Reports of the Research Committee, No. XL, 1982, pp. 95–8, Fig. 1, Fig. 2.

7. Excavated by Newbury and Grove in 1975/6. There are no published records of this archaeological work. I am grateful to Mr D.B. Kelly of Maidstone Museum and Art Gallery for his observations and dating of the sherds.

8. Including, Holeywell Hill, see *Journal of Roman Studies*, vol. LIX, 1969, p. 221; Heronbridge, Cheshire, see T. Garlick, *Romans in Cheshire*, 1973, p. 52; B.R. Hartley, 'Excavations at Heronbridge 1947–48' in *Chester Archaeological Society Journal*, Vol. 39, pp. 1–21; B.R. Hartley and K.F. Kaine, 'Roman Dock and Buildings' in *Chester Archaeological Society Journal*, Vol. 41, pp. 15–38: Kimpton, Hampshire, see R. Goodburn, 'Roman Britain in 1978' in *Britannia*, Vol. x, 1979, p. 331; Hampshire Field Club, *Newsletter*, Vol. 4, Sept. 1976, p. 9: Littlecote Park, Wiltshire, see *Britannia*, Vol. xiii, 1982, pp. 387–8; *idem*, Vol. xii, 1981, p. 360 + Fig. 16; *idem*, Vol. x, 1979, p. 329 + Fig. 17: Kenchester, Herefordshire, see *Britannia*, Vol. ix, 1978, p. 438; *idem*, Vol. ix, 1979, p. 298; *West Midlands Archaeological News Sheet*, Vol. 20, 1977, pp. 33–6; *idem*, Vol. 21, 1978, pp. 69ff. I am grateful to Mr B. Phillips for bringing the Kenchester site to my attention.

9. C.J. Young, 'Excavations at Ickham' in *Archaeologia Cantiana*, Vol. xci, 1975, pp. 190–1; C.J. Young, 'The Late Roman Water-Mill at Ickham, Kent, and the Saxon Shore' in *Collectanea Historica, Essays in Memory of Stuart Rigold*, Maidstone, 1981, pp. 32–9.

10. Including Dr J.D. Ogilvie, A. Burbridge, K. Elkes, P. McWhirter and Lady Empson.

11. *Archaeologia Cantiana*, Vol. lxxiii, 1959, p. lii; *idem*, Vol. lxxii, 1958, p. lxiv.

12. Ivan D. Margary, *Roman Ways in the Weald*, London, 1948, reprinted 1968, pp. 15–21.

13. *Transactions Cumberland and Westmorland Archaeological Society*, 5, Vol. 13, 1913, p. 369.

14. A practice met with in Germany; see C.F. Innocent, *The Development of English Building Construction*, 1916, reprinted 1971, p. 110, quoting Moritz Heyne, *Deutsche Wohungswesen*, p. 18.

15. The mill site was dug away in 1974 by a floating dredger and at one point the archaeological work was conducted from a boat moored in 30 ft of water.

16. Vitruvius, *De Architectura*, Loeb Classical Library, 2 vols., London, 1962. The relevant passages are in Book X, Chapters IV and V.

17. The axes of a single quadrilateral sectioned post are referred to as minor; a major axis is an axis common to more than one post.

18. This is Vitruvius' vertical gear wheel, the *tympanum dentatum ad perpendiculum conlocatum*, a face gear with either protruding pegs or cog-like teeth. In modern mills the equivalent gear is the pit-wheel, a gear which as the name implies normally ran in a pit. In discussing this watermill it would be misleading to use this term because there may not have been a pit as such. For clarity, it is suggested that the self-explanatory names of 'driver' and 'driven' gear should be used.

19. Some archaeologists have referred to the shaft which supports and drives the millstones as a mill-pivot or pivot. The word pivot is mis-applied because there is a power transmission function. In watermill terminology this is referred to as the millstone shaft or stone-shaft. In earlier watermills, including all known Roman examples, the millstone shafts are much smaller in section than their modern counterparts and, in keeping with general engineering practice, it will be called a millstone spindle or, for brevity, spindle.

20. Of this there can be little doubt; iron was a far too valuable commodity

to be used for a shaft 2·5 m long, especially where a hardwood could serve the function just as well.

21. The axle of the Rio Tinto drainage wheel at the British Museum (Reg. No. 1889 6-22.1) was described as being of copper on the acquisition of the object but this has not been metallurgically proven. Other Rio Tinto wheels had bronze axles; R.E. Palmer, 'Notes on some Ancient Mine Equipments and Systems' in *Transactions Institute Mining and Metallurgy*, Vol. xxvi, 1926-7, p. 256.

22. This figure is based on experiments conducted by many of the early water-power engineers who attempted to establish the optimum velocity for the floats of an undershot waterwheel. Parent, Pitot, Desaguliers, Maclaurin and Ferguson all adopted one-third as the relationship between the float velocity and water velocity; see *Ferguson's Lectures on Select Subjects* with notes and appendix by David Brewster, second edition in 2 vols., Vol. i, 1806, Edinburgh, p. 168. However, greater reliance is placed on the experiments of John Smeaton who considered the optimum figure to be nearer one-half than one-third. The mathematically derived ratio is 1 : 2.

23. A good example is Click Mill at Dounby in the Orkney Islands which is maintained by the Department of the Environment. Here the meal passes through a short wooden spout and is collected in a bin installed in front of the bed-stone.

24. For safety reasons the main gearing in modern watermills is invariably enclosed in a chamber created by boarding on the frame and/or separate stud walls. At Ickham the machinery may not have been divided off but the number of surrounding posts and beams would certainly have given the impression of an enclosure. The word chamber is used throughout because it is conveniently definitive.

25. When the grain is being ground, the weight of the upper millstone is not taken by the spindle but is instead supported by the lower millstone. Indeed, there should always be grain between the stones or else they will rapidly become defaced.

26. I.A. Richmond, *Roman Timber Building: Studies in Building History*, O'Niel memorial volume, edited by E.M. Jope, 1961, p. 18.

27. This is determined by two considerations, satisfactory access for a man and clearance for the gears above the wheel-shaft.

28. The brayer is a pivoted beam which lifts one end of the bridge-tree in order to adjust the gap between the millstones.

29. In mill terminology the word 'trough' is normally applied to the man-made U-shaped wooden or cast-iron channel found at the mill end of the head-race where it delivers water onto the wheel. The term is especially applicable to breast and overshot wheels. In the text the application of the word is nearer to its general meaning.

30. L.F. Salzman, *Building in England Down to 1450*, 1952, p. 223, quoting Bede, *Opera*, ed. Plummer, Vol. i, p. 147.

31. Five Roman feet; 1·48 m; 4·86 ft.

32. Arthur W. Parsons, 'A Roman Water-Mill in the Athenian Agora' in *Hesperia, Journal of the American School of Classical Studies of Athens*, Vol. V, No. 1, 1936, pp. 70-90.

33. F. Benoit, 'L'usine de Meunerie Hydraulique de Barbegal (Arles)' in *Revue Archeologique*, series 6, Vol. 15, 1940, pp. 19-80, Fig. 3.

34. The clue comes from post 2 which appears to have been a young tree. This would have contained a high proportion of sapwood and probably decayed before any other.

35. The identification and interpretation of the coins is contained in a report submitted to the author by Mr K.G. Elkes.

36. C.J. Young, as note 9.

37. R.J. Spain, An Analysis of the Millstone and Quern Fragments from Ickham, 1980, unpublished report.

38. Two iron spindles were found at the Roman fort of Zugmantel on the Domitian *limes*. One with a gear still attached was recovered from the well of one of the *vicus* houses together with two millstones. It is thought that these mill parts were thrown down the well in the second half of the second century AD. See H. Jacobi, 'Romische Getreidemuhlen' in *Saalburg-Jahrbuch*, Vol. 3, 1912, pp. 75-95, 89, Fig. 43. Also 'Kastell Zugmantel, Die Ausgrabungen' in *Saalburg-Jahrbuch*, Vol. 3, 1912, p. 54, Figs. 17, 18.

39. *Idem.*, Taf XVII. The reconstructed mill is in the Saalburg Museum.

40. Parsons, as note 32.

41. Coin evidence suggests a date for construction during the reign of Leo I (AD 457-74) and a destruction in the reign of Justin II (AD 565-78) or not long after.

42. In view of the fact that metal ferrules or hoops were found at each end of the shaft, it seems more likely that the journals were of iron or bronze. Iron on wood has a better coefficient of friction and longer life than wood on wood. If the journal was of wood, it would have only lasted a short time before the millers were faced with either shaft removal, which necessitated dismounting the wheel and gear, or far easier, boring the shaft ends and inserting metal journals.

43. A rather clumsy arrangement which would have required wedging to the walls of the pit to give stability. Parsons, as note 32, considered that the block of marble below the east bearing was indicative of this arrangement. A more stable and easier arrangement would have been for the bridge-tree to span across the pit and shaft, perhaps tied to the sill beams of the hursting.

44. Parsons, as note 32, p. 85.

45. T. Schiøler and O. Wikander, 'A Roman Water-Mill in the Baths of Caracalla' in *Opuscula Romana XIV:4*, Skrifter Utgivna Av Svenska Institutet I Rom, 4, XXXIX, Stockholm, 1979.

46. There were two independent waterwheels each driving a single pair of millstones.

47. Schiøler and Wikander, as note 45, p. 54 and Fig. 12. The authors imply that a traditional box-frame hursting, integral with the bridge-tree existed, but this is most unlikely.

48. The three watermills are Barbegal, the Athenian Agora and the Baths of Caracalla.

49. C. Reindl, 'Ein romisches Wasserrad' in *Wasserkraft und Wasserwirtschaft*, Heft. 11/12, 34 Jahrg., 1939, pp. 142-3; L. Jacono, 'La ruota idraulica di Venafro' in *L'ingegnere*, Vol. 12, 15 Dec. 1938, pp. 850-3. A model of this wheel is in the Technological Section of Naples Museum.

50. The following sites have provided such evidence: Barbegal, Athenian Agora, Baths of Caracalla and Haltwhistle Burn Head.

51. O. Davies, *Roman Mines in Europe*, 1935, p. 26; Stevenson, *Archaeologia Aeliana*, Vol. VII, 1866, pp. 279-81.

52. R.E. Palmer, *Transactions Institute Mining and Metallurgy*, Vol. XXXVI, 1926-7, pp. 299-310; R.J. Forbes, *Studies in Ancient Technology*, Vol. VII, Leiden, 1963, pp. 212-17; Davies, as note 51. One of the Rio Tinto drainage wheels is in the Department of Greek and Roman Antiquities of the British Museum, Reg. No. 1889. 6-22.1.

53. F. Posepny, *Mitt. authr. Ges. Wien.*, Vol. XII, 1892, p. 44; *id. Ost. Zeitschr. F. Berg-u Huttenwesen*, Vol. XVI, 1868, pp. 153–4, 165–8; Vol. XXV, 1878, pp. 391–3 (Verespatak); see also G.C. Boon and C. Williams, 'The Dolaucothi Drainage Wheel' in *Journal of Roman Studies*, Vol. LVI, 1966, pp. 122–7.

54. *The Great Palace of the Byzantium Emperors*, 1947, p. 83, plates 41–2. Report on the excavations carried out on behalf of the Walker Trust (University of St Andrews), 1935–8.

55. O. Wikander, 'Water-Mills in Ancient Rome' in *Opuscula Romana*, Vol. XII: 2; *Skrifter Utgivna Av Svenska Institutet I Rom*, 4, XXXVI: 2, Stockholm, 1979, p. 23, Fig. 11; T. Schiøler, *Roman and Islamic Water-Lifting Wheels*, Odense, 1973, pp. 154–5, Fig. 110.

56. Refer to note 38.

The Gas Engine in British Agriculture *c.* 1870–1925

IAN R. WINSHIP

Introduction

A study of the use of gas engines on farms might not seem particularly significant, but there are two reasons why such a study is useful. Specifically, the gas engine was the first widely available source of mechanical power available to farmers and hence represents an important stage in the mechanization of farm work. More generally, a study of agricultural uses demonstrates how incorrect is the frequently met attitude that gas engines were merely a step on the way to the modern internal combustion engine. Histories of power tend to discuss the years leading up to Otto's 4-stroke engine of 1876 and then pass on to the petrol engine of the next decade as if the gas engine was merely a short-lived device of little application. Yet, as Barlow, in his very welcome PhD thesis reminds us, nearly 27,000 Otto engines had been built by the mid-1880s and were in use in dozens of different ways in industry, dairies, offices, printers, sawmills and for organ blowing, refrigeration, electricity generation, and so on.[1] Thus the gas engine was by no means a passing phase of little or no consequence.

This short paper examines the application of the gas engine to agriculture by looking at the extent of its use, the tasks to which it was applied and the economic and other advantages it provided. Unfortunately, a shortage of readily available source material means that on some points it is difficult to be conclusive. There seems to be very little published on the details of the gas engine's use especially before the present century, with most material dealing with what was available rather than discussing which engines were actually used. Similarly the earliest statistics available are for 1908 when the gas engine was no longer an important source of power. Accordingly this account is presented not as a comprehensive history but as an exploration of the nature of the problem providing a basis for further work.

STATISTICAL NOTE

There are a number of ways of quantifying the output of engines. Initially with steam engines it was a theoretical value based on cylinder dimensions and boiler pressure and quoted as nhp (nominal horsepower) or more likely just as hp. This method was inaccurate, especially for large outputs and for internal combustion engines, so that more precise ways were devised based on measurement of output, namely indicated horsepower (ihp) as determined in the cylinder and effective or brake horsepower (bhp) determined at the driveshaft. Bhp is the more accurate measure of available power and is only three-quarters or so the value of ihp because of transmission losses. Nhp is about half the value of ihp for engines up to around 20 ihp after which it is a decreasing proportion so that, for example, 100 ihp is only about 35 nhp. Where possible, bhp is used in this paper but many published figures, especially in the official statistics, merely say hp, that is, nhp.

Early Use of Power on Farms

Before the 1870s, mechanical power on farms was limited in both extent and in sources available. Water power was in use though our detailed knowledge of its application in agriculture is sparse. There were obvious topographical limitations to its adoption and though we may be surprised that it lingered well into the twentieth century, with 3,800 wheels in England and Wales in 1925[2] (a similar total to steam engines then remaining), it is not surprising that most were in mountainous Wales. What is especially interesting is that it could still be claimed in 1906 that 'where water power is available most farmers think no source is equal to it for driving fixed machinery'.[3] However, it seems clear that despite developments in its technology, including the use of turbines, water power did not play a great part in power provision. The same is true for wind power. Of greater importance, of course, was steam power. The first stationary engine was installed by the iron founder John Wilkinson on his Denbighshire farm in 1798,[4] and we may assume it was a Boulton and Watt type. This, like most others installed later, was used to drive threshing machines though some were used for driving stationary barn machinery as well, the earliest being noted in 1804. Fixed steam engines were never especially successful because the power needs were not great enough to justify the expense and labour necessary. Steam power proved far more useful when it was mobile, either in the field ploughing by cable or as portable traction engines for threshers. A farmer would rather hire an engine from a contractor for the few days it took to thresh his crop than own one that would be underused. And of course cost was important. In 1853 a 10 hp engine could cost £350 to which might be added £100 or so for the building to accommodate it, even before considering its running costs.[5] When at around the same time a hand-operated chaff cutter cost about £6, a turnip cutter around £4[6] and a labourer's wage varied in different parts of the country from 11–16s/week,[7] then clearly it was not economic to dedicate engines to this sort of work and only on the largest farms where threshing would provide most of the work for the engine would other machinery be mechanically powered.

The ownership (if not the use) of steam engines was never really widespread. It was calculated in 1860 that 10,000 hp/year was being added to agricultural power[8] but that seems doubtful when the Agricultural Census of 1908 gives only 106,000 in total.[9] Admittedly steam was past its peak then but it is not likely to have declined very much. Other sources of power comprised a similar amount and are likely to represent to a large extent additional power provision rather than replacement, so that we might assume this figure to be of a similar order to that at the peak of steam use. Of more interest is the total number of engines in 1908 (17,000 and probably an overestimate) for this represents 1 for every 30 farms of an acre or more or 1 for every 10 farms over 50 acres. Though portable engines would extend power to a greater number of farms, it seems likely that power, for fixed barn machinery before the advent of the internal combustion engine (and even for many years after), was provided mainly by horse and man.

The Gas Engine

DEVELOPMENT

A brief outline of the major developments in the gas engine is necessary to see its use in agriculture against the general availability of the engine.[10] Development can conveniently be divided into three periods—that up to Otto's patent for his 4-stroke engine in 1876; that during the patent's duration till 1890; and the years after. In the first period came the earliest successful production engine of Lenoir of which a few hundred were built from 1860 but none are known to have been used in agriculture. Of greater importance are the engines of Otto and Langen and Bisschop. Otto and Langen's successful joint engine was that of 1867—a vertical atmospheric engine in which the 'free' piston was forced upwards by the combustion of a gas and air mixture. On the down stroke the piston engaged via a frictional clutch with a toothed wheel attached to a fly-wheel. About 80 rpm was achieved with an output up to 2 bhp. The Bisschop engine was constructed in 1870 and, despite the Otto–Langen engine having established itself, found a market for very low power. It had some similarities in principle to the Otto–Langen though the piston was not free and it dispensed with water cooling. Speed was higher but output lower—from 1 to 4 manpower. It was compact, simply constructed and required no lubrication of the cylinder. It could be 'managed by any boy or girl'. Some thousands were built over a quarter of a century or so, especially by J.E.H. Andrew of Reddish who in 1881 charged £25 for the 1 manpower model ($c. \frac{1}{2}$ bhp) and claimed the cost of gas consumption as $\frac{1}{2}$d/hour.

Otto's 4-stroke engine of 1876, despite its physical similarity to a horizontal steam engine, marked the beginnings of the modern internal combustion engine. Just as the introduction of higher steam pressures in Cornish engines had given great impetus to the development of the steam engine, so Otto's use of compression of the gas/air mixture similarly stimulated acceptance of the gas engine. Gas supply was controlled by a slide valve which incorporated a gas burner to ignite the charge. These

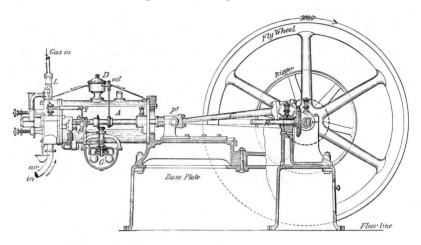

Figure 1.   Otto's engine, 1876 (*from Donkin, as note 10*).

valves were prone to wear and required frequent maintenance so were replaced in 1885 by Crossley Brothers, the UK licensees, with poppet lifting valves and hot tube ignition. With the latter, which became a method generally used, a cast-iron tube was kept red hot by a bunsen burner. The compressed charge was introduced into the tube and was ignited. Otto's engine was practical, far superior in performance to its predecessors, lighter in weight, had improved thermal and mechanical efficiency and was economical in use. Not surprisingly it found a ready market. By 1887 some 27,000 had been sold worldwide, built by Otto's company Gasmotoren Fabrik Deutz, by Crossley Brothers in Manchester and by other companies. Crossleys, who had built the Otto–Langen engine, were soon producing the new engine, their first one being exhibited at the Royal Agricultural Society Show in Liverpool in 1877. Whether this was in anticipation of sales to farmers—because conveniently they had an agent in Liverpool—or simply that this was the first available public exhibition they could use is not clear. Only two years later, Crossleys were offering a range of eight engines from $\frac{1}{2}$ nhp at £60 to 16 nhp for £350.

Because of Otto's success and the limitations his patent imposed on others (mainly through its successful defence in court) there was little significant development over the next decade. What there was concerned 2-stroke engines though none achieved great success. James Robson, who had produced a small 2-manpower vertical atmospheric engine exhibited at the Smithfield Club Cattle Show in London in 1880,[11] soon after developed a 2-stroke horizontal compression engine that was built by Tangye of Birmingham from 1881–90. In principle, this anticipated the modern 2-stroke and was ignited by a slide valve burner. Some hundreds were built. Robson's rival in the development of the 2-stroke engine was Dugald Clerk who claimed to have produced the first successful engine with

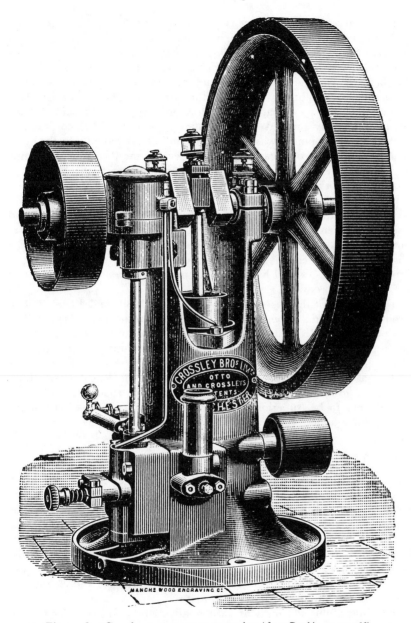

Figure 2. Crossley two-manpower engine (*from Donkin, as note 10*).

compression and ignition at every revolution. His engine was introduced in 1878 and after testing was first exhibited at the Royal Agricultural Society's Show in London in 1879. However, it never went into production and it was an improved version that was built by Thomson, Sterne and later by Tangye. Only a few hundred were built, perhaps because of a low efficiency and the ever-present threat of infringement suits by Otto.

Rather in the way steam engine development might be said to have been retarded by James Watt's monopoly from 1769–1800, so practical gas engine development was hindered during Otto's patent. On the other hand, the various court cases tended to develop the theory of the engine as witnesses tried to show how their designs differed from Otto's or to analyse the workings of the various engines. Accordingly, after the patent expired in 1890 there was a great variety of 4-stroke engines produced. None showed the significant advances over earlier engines that Otto's had and developments were more concerned with the search for higher power outputs—up to 300 bhp with twin cylinders; with the use of high-speed engines to drive electricity generators; and the use of self-contained gas-producing plant. Table 1 lists the major manufacturers of gas engines and the types they produced.

The heyday of the gas engine is seen as the 1880s and 1890s, though production continued long after that. Crossley Brothers alone sold 40,000 engines between 1877 and 1900 (with a total ihp of half a million) and the annual growth in sales was only then beginning to show signs of slackening. (In 1900, over 3,000 engines were produced.) Even in the 1920s gas engines were still being exhibited at the Royal Agricultural Society's shows and an agricultural machinery textbook of 1923 could claim them to be 'superior for stationary machinery when a town gas supply is available'.[12]

FUEL SUPPLY[13]

The first gas engines—probably including the first used on farms—were fuelled by ordinary town gas. The provision of a gas-supply network began in London in 1814 and soon spread to other areas beginning, for example, in Newcastle in 1818 and Morpeth in 1833. By mid-century, gas lighting was well established in urban and semi-rural areas. It was never extended to remote rural habitation in the way electricity was in the present century. Accordingly, the only farms that would have a mains gas supply would be those on the edge of towns and villages and clearly this was a limitation to the use of the gas engine.

This inflexibility of supply and the relatively high fuel costs for engines above *c.* 30 nhp soon led to the provision of small self-contained oil or coal-fired gas-producing plant. Mansfield's oil gas system (shown, for example, at the Royal Agricultural Society's Show in 1897 but developed much earlier) could use various vegetable oils, animal fats, etc. but usually employed shale oil. In a process similar to distillation from coal, a gas twice as rich as lighting gas was produced (typical calorific values were,

TABLE 1   Gas engine manufacture in Great Britain, 1870–1900 (data taken from Barlow, note 1 and Donkin, note 10)

Manufacturer	Place of Production	Trade Name of Engine or Basic Type	Main Technical Features
Alexander, Burt	Glasgow	'Acme'	2 cylinder, 2 pistons, 2 crankshafts
J.E.H. Andrew	Reddish	Bisschop's 'Stockport'	1870 atmospheric 1882 4 stroke
T.B. Barker	Birmingham	'Forward'	1890s 4 stroke
Campbell Gas Engine Co.	Halifax		2 stroke, 4 stroke
Clarke Chapman & Co.	Gateshead		rotary valves, hot tube or electric ignition
Crossley Brothers	Manchester	Otto and Langen's Otto's	c. 1870 1877 4 stroke 1886 vertical 1890s twin cylinder
Day & Co.	Bath		vertical 2 stroke, 1 or 2 cylinders
Dick, Kerr & Co.	Kilmarnock	Griffin's	1880s 3 cycle
Fawcett, Preston & Co.	Liverpool		2 cylinder, similar to Clerk's
Fielding and Platt	Gloucester		1891 similar to Otto
Furnival	Reddish	'Express'	Otto type
Gardner	Manchester		1894 good workmanship, bench tested
Grice & Son	Birmingham	'Birmingham'	for small workshops
Hindle & Norton	Oldham	Dougill's	small 4 stroke
National Gas Engine Co.	Ashton under Lyne		1890 Otto type
Palatine Engine Co.	Liverpool		1890s, vertical 2 stroke, high speed
Robey & Co.	Lincoln		1890 especially for electricity generation
A.E. & H. Robinson	Manchester		4 stroke, small, high speed
John Robson	Shipley	'Nonpareil' 'Shipley'	Otto type
Simon	Nottingham	Brayton's	1878 vertical, separate pump
Tangye	Birmingham	Robson's	1881 2 stroke vertical 1891 large engines, mostly with gas plant

TABLE 1 (*cont*)

Manufacturer	Place of Production	Trade Name of Engine or Basic Type	Main Technical Features
Taylor	Nottingham	'Midland'	before 1890 motor and pump system 1890s 4 stroke
Thompson, Sterne	Glasgow	Clerk's	1881 2 stroke
Trent Gas Engine Co.	Nottingham		1891 2 cylinder tandem compound
Wells Bros.	Nottingham	'Premier'	4 stroke, vertical and horizontal
Weyman	Guildford	'Trusty'	often 2 or 3 cylinders side by side, 4 stroke

respectively, 480 and 230 BTU/ft³) for about 6d/100 ft³, which Donkin claims is 'much more expensive than coal gas in England'. Yet though coal gas was around 2s 6d to 4s/1,000 ft³, or 3 to 5d/100 ft³, twice the consumption of oil gas was needed: Donkin quotes Crossley figures of 10 ft³/bhp/hr for the latter to compare with the usual range of 17–25 ft³ for coal gas.[14] Thus oil gas would cost 0·6d/bhp/hr and coal gas 0·5 to 1·2d/bhp/hr. Consequently a self-contained gas plant might have advantages for large engines used intensively or where wastes could be used as fuel, which would not seem to be the case in agriculture.

The most widely used plant in the first twenty years or so after Otto's engine was Dowson's coal-based process developed in 1878–9. In this process, steam and air were passed through a generator containing heated coke or anthracite. The combustion produced gases which were cleaned in various ways before passing to the engine.[15] The gas was cheaper than other types and could cost as little as 0·1d/hp/hr, though fuel costs seemed variable. Anthracite and gas-coke were expensive in Britain in the 1880s, ranging from 7s 6d per ton to 2 or 3 times that amount. Nevertheless gas costs were very low. The cost of plant had to be added, of course, and in 1888 varied from £125 for plant suitable for a 7 ihp engine to £205 for a 60 ihp engine. There were disadvantages in that the gas partially comprised poisonous carbon monoxide and the heating value was low (only 135 BTU/ft³) so that it could not be used for flame ignition in slide valve engines, though it was quite suitable for tube ignition. Its combustion properties were different from coal gas so some engine modifications were required to allow a different gas/air mixture. A higher compression ratio was needed too, which helped to improve engine performance. Again, use in agriculture seems doubtful unless large engines were required. Crossleys' literature of 1888 on uses in corn grinding and similar work[16] includes a few engines with Dowson plant but they were all 16 hp engines in corn mills. The farm engines listed were mostly of 2–6 nhp. An 1896 review of farm motive power notes that coal gas was used where obtainable, otherwise a gas was manufactured, most economically by Dowson's process.[17] However, it is not clear if these remarks refer to agriculture or to gas engine practice in general—the latter may well be the case.

Of more relevance to agriculture was the suction gas plant introduced around 1900. The process was an improvement on Dowson's in that air was sucked through the generator by the induction stroke of the engine. According to Barlow the gas was filtered to prevent particles of grit or ash being drawn into the engine but no washing or scrubbing of the gas took place. However, contemporary literature[18] makes it clear that coke-filled scrubbers were normal practice and indeed dominated the plant as Fig. 3 shows. The plant was more compact than Dowson's and portable versions were available. The gas produced had a calorific value similar to that from Dowson's process but the cost was lower—about 0·05d/bhp/hr—and the process could be fuelled by coal, coke or anthracite. Coal and coke were cheaper but gave rise to tarry compounds in the gas, especially with coal, so necessitating more elaborate cleaning devices. A National Gas Co. plant of 1922 was reported to burn the tar, though it was not explained how.[19] Some plant could also make use of wood waste, sawdust or peat.[20] It was thought that the residue in the generator might have manurial value but the Royal Agricultural Society's consultant chemist showed it was comprised almost completely of moisture and carbonaceous matter with only very small quantities of nitrogen and trace elements.[21] Interest in suction gas plant by the agricultural community was such that the Highland and Agricultural Society of Scotland held trials in 1905 and the Royal Agricultural Society of England in the following year. The Scottish trials involving 10 hours' running were conclusive—'economy, efficiency and simplicity of working of the suction gas producer plant was demonstrated beyond question'.[22] The aims of the Royal's more extensive trials were to see if suction gas plant could work for agricultural purpose day in

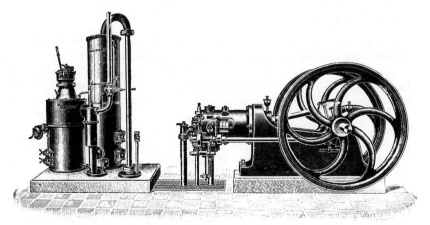

Figure 3. National Gas Company 20 hp engine and suction plant (from H.E. Wimperis, *The Internal Combustion Engine*, Constable, 1915).

day out with the same freedom from breakdown as steam engines and with the same small amount of attendance as oil engines.[23] Their improved fuel economy over steam and oil engines was not in question. Results fully justified the organizers' hopes and a gold medal was awarded to a 20 bhp plant from the National Gas Co. (Fig. 3 shows a similar plant). The report concludes: 'It is probable that suction gas plants will be able to replace portable steam engines for many purposes with distinct advantages as regards handiness and economy.' Both trials recorded similar consumptions with, on average, a little over 1 lb of coal or coke/bhp/hr compared with a figure (admittedly from 10–15 years earlier) for Dowson plant of 1·5 lb.

Suction gas plants were subsequently prominent at the Royal Agricultural Society's shows, usually being offered with an engine as an integrated unit. For example, virtually every engine at the 1908 show at Newcastle was exhibited with a gas plant.[24] It is difficult, however, to assess how widespread was the practical use of such plant. A writer in 1906, while noting that suction gas plant provided the cheapest power ever, admits that 'as far as has been learned no installation of suction gas producer is yet erected on any farm but numerous enquiries have been made ... their successful introduction on farms is only a matter of time'.[25] Moreover, a textbook of 1923[26] notes that suction plant were applicable only for comparatively large power requirements, a judgement that is reflected in a closer examination of the exhibits at the RASE shows where the engines with plant are of at least 20 bhp. Given that the average size of gas engines in agriculture in 1908 was 5 hp,[27] one would expect the number of engines of 20 or 30 hp or more to be small. Further, it seems doubtful that many new engines would be introduced on farms after this as oil engines were by then the major source of power.

Gas Engines on Farms

SUITABILITY

The gas engine's use in agriculture is of note not because the engine became commonplace but because it represents the start of the widespread application of mechanical power. The figures for the 1908 Agricultural Census (Table 2) show that only then were internal combustion engines of all types beginning to equal the number of steam engines in use. Gas engines represented only a small proportion of the available motive power—5 per cent by number and 4 per cent by horsepower. To some extent this is surprising given the apparent enthusiasm in the early years— a report on the exhibits at the RAS show of 1887 comments that the petrol engine is 'a means of obtaining power without a steam engine, an advantage which the enormous sale of gas engines proves to be both great and general'.[28] On the other hand, the same writer four years later in discussing agricultural prime movers made no mention of gas engines at all![29]

The advantages offered by the gas engine were many. Crossley Brothers' literature[30] of 1888 listed numerous advantages over the steam engine— there was no tedious preparation before starting; very little cleaning was needed; no chimney to build; no smoke nuisance; there was no coal to get in or ashes to cart away (though this was no longer true, of course, if producer plant were used); no dust; no separate boiler and the consequent risk of explosion; no gauges or valves to watch; no fire to stack down or draw at stopping time; no regular attendance beyond oiling, cleaning and starting; great economy. Donkin, a few years later,[31] added that the gas engine was simpler because it was complete in itself; it was compact; it could be fixed almost anywhere, but should stand on a solid foundation; installation was easy; it was stopped simply by turning off the gas supply; there was less danger of fire so it could be used in places where steam could never be employed. He suggested, however, that small gas engines were more costly than steam but noted that efficiency was always improving. Around the same time a survey of agricultural power[32] notes that gas (and oil) engines require little skill in management and have fewer expensive parts such as boiler and firebox to go wrong. A disadvantage was that there was no steam or hot water to use for cleaning, heating or cooking animal fodder. A later description[33] notes the engine to be clean and reliable, comparatively free from troubles associated with incomplete combustion, and economical. This appraisal, however, was as likely to be in relation to oil engines as to steam; though it is interesting that in the early 1920s the gas engine was still a power option to be considered seriously. In essence the advantages the gas engine could bring to agriculture were as a compact source of power that could be used intermittently, was easily started and stopped and required minimum attention in operation.

The major disadvantage was the need for a piped gas supply or the extra expense and work of installing a gas producer plant. The oil engine quite naturally was promoted as overcoming such problems. Thus in the same year as Crossleys were extolling the gas engine, Priestmans were advertising their petrol engine with emphasis on its freedom from the very costly connections needed for gas, the absence of slide valves, self-lubrication and simplicity of construction.[34] Donkin[35] suggested that any engine that would overcome the drawbacks of steam and gas engines must be self-contained and quite independent; safe and simple, using as a working agent a fuel which was neither difficult to obtain nor dangerous to transport; easy to handle so any unskilled workman could drive it; compact and easily transported from place to place; and economical in working. 'These conditions are found in the Priestman engine which is well adapted for all kinds of industrial operations requiring small powers. In many country places where gas cannot be procured ... it has probably a great future before it.'

EXTENT OF USE

Considering some of the above remarks, it is perhaps surprising that the gas engine did not disappear from use after only a few years. A survey of gas companies in the UK in January 1897[36] showed that 25,700 engines were being run from public supplies and as has already been noted Crossley Brothers were still increasing their output during the 1890s. However, by comparison with oil and petrol engines this figure may have been declining or growing less rapidly if the figures for agriculture of a few years later are any indication (Table 2).

Unfortunately we have no detailed figures of agricultural power use before these of 1908. For earlier years we have mere indications of the use of gas engines from manufacturers' literature. Thus the American agent for the Otto–Langen engine notes in an 1876 advertisement[37] that more than 3,000 engines had been made and the list of applications includes 80 as 'agricultural machines'. As some 30 per cent of the engines were made by Crossley Brothers, it seems likely that some of the agricultural machines would have been used in Britain. Similarly an 1881 advertisement by J.E.H. Andrew for the Bisschop engine[38] claims that over 1,000 engines were in use in the UK for varied applications including chaff cutters (this appears third in the list after printing machines and pumps which may be significant). An 1887 price list for the Bisschop and Andrew's own 'Stockport' engine[39] includes 'chaff cutters, corn crushers, churns, etc' among applications as well as driving saws, which could be an agricultural use.

The most extensive available listing of applications of the gas engine is the collection of testimonials from users and lists of users published by Crossley Brothers around 1888 and reproduced by Barlow in his thesis.[40] One section is entitled 'Selected list of users ... driving appliances in

TABLE 2   Motive power on farms in Great Britain, 1908

| | | Size Group of Holdings (acres) | | | | |
		1–5	5–50	50–300	300+	Total
Holdings		108,094	231,819	151,002	17,714	508,629
Steam	no.	249	1,595	9,160	5,955	16,959
	hp	1,576	8,010	53,330	43,544	106,460
Oil	no.	118	1,199	8,604	2,886	12,807
	hp	670	5,747	55,307	22,516	84,240
Petrol	no.	18	155	919	255	1,347
	hp	48	478	3,387	1,128	5,041
Gas	no.	131	670	897	157	1,855
	hp	435	2,591	4,967	1,511	9,504
Others	no.	29	211	854	388	1,482
	hp	107	828	4,459	2,886	8,280
Total	no.	545	3,830	20,434	9,641	34,450
	hp	2,836	17,654	121,450	71,585	213,525

(Data from Board of Agriculture and Fisheries, as note 9.)

connection with corn grinding, fodder chopping and similar work only' and lists some 634 engines of $\frac{1}{2}$ to 16 hp. However, examination shows that most were used for animal food preparation for tramway companies, stables, co-operative societies, agricultural suppliers such as the Northumberland Agricultural Association, Alnwick, and so on and only sixteen can be definitely identified as being on farms. However, the addresses given do not always clearly indicate farms and uses such as 'general work'; 'estate work' could mask agricultural applications. Moreover these lists are selective and clearly do not include all the 27,000 engines that Crossleys had built by then. Those engines listed ranged geographically from Belfast to Cardiff to Sussex to North Yorkshire and in engine size from 1 to 6 horsepower.

The official figures available are those from the Agricultural Censuses of 1908, 1913 and 1925. Those published in the report of the 1908 census are shown in Table 2 and those from the report of the 1925 census in Table 3. (Because of the war there was no separate report of the 1913 census.) What is immediately apparent is the great difference between the two sets of figures for 1908; one that is unlikely to be accounted for by the different geographical coverage. According to Britton and Keith's survey of power supply statistics,[41] the 1908 report made generous additions for engines owned by farmers who did not answer the supplementary question on machines. This manipulation is not admitted in the 1908 report but it is pointed out that only about 80 per cent of returns included replies to the supplementary questions and as the total of holdings in Table 2 is the total known rather than returns made, then other figures may have been increased too. The 1908 figures in Table 3 are the actual ones returned, as are the others. The 1925 report notes that returns that year were received from only 59 per cent of the acreage of the country and estimates that the figures cover only 70–80 per cent of engines in use.

Though the absolute figures in Table 2 are doubtful, it seems reasonable

TABLE 3   Agricultural engines in use in England and Wales, 1908, 1913 and 1925

	1908	*1913*	*1925*
Fixed or portable			
Steam	8,690	7,719	3,731
Gas	921	1,287	1,125
Oil or petrol	6,911	16,284	56,744
Electric	146	262	700
Wind ⎱	—	3,663	990
Water ⎰			3,543
Others not stated	663	28	21
Motor tractors			
For field operations	—	—	14,565
Solely for stationary work	—	—	2,116

(Data from Ministry of Agriculture and Fisheries, as note 2.)

to assume that the distribution of power of different types and on different sizes of farm is broadly correct. Thus we can see that in numerical terms, 49 per cent of engines were steam, 37 per cent oil and only 5 per cent gas. In terms of horsepower, the proportions are almost the same: 50, 39 and 4. The average power of engine was 5·9 hp for steam, 6·6 hp for oil and 5·1 hp for gas. As might be expected the larger farms had the more powerful engines but the differences were more extreme for gas engines. For steam engines the average power for the four sizes of holding were 6, 5·3, 6 and 7·2 hp; for gas the power was 3·3, 3·9, 5·5 and 9·6 hp respectively. The lower figure is easily explained from the widespread use of low-rated engines but the higher one is more puzzling. However, given the small number of engines on these largest farms, then only a handful of engines of high power would be needed to inflate this average. More generally, we can see how little mechanical power there was employed at this time—only 7 per cent of farms had any kind of engine. Thus the great increase in power shown by the growth of oil engines and tractors (Table 3) is a result of previously unmechanized farms installing machines. It seems less likely that there was any widespread replacement of other power sources while they still performed satisfactorily. So the number of steam engines in this period did not really decline until after the War and the number of gas engines remained relatively static. Indeed, given that the 1925 figures are admitted to be more of an underestimate than the other, then gas engines may even have increased in number by 1925, though obviously in proportional terms they were becoming even less significant.

The 1925 report indicates for that year the regional distribution of the different types of engine. By presenting results in terms of the number of engines/10,000 acres of crops and grass, the report also shows how few engines of some types there were, with a ratio of only 0·74 for gas engines compared with 37·2 for oil and petrol. Gas engines were most numerous in the Northern Division (Northumberland, Durham, North and West Ridings) with 2·06, closely followed by North Western Division (Cumberland, Westmorland, Lancashire, Cheshire, Staffs, Derbyshire). Least use was in Wales and the south west, the areas of greatest use for oil engines. The report suggests that gas was more popular in the north because of more readily available rural supplies of gas and that oil and petrol engines were more numerous in pasture than arable counties for their chief use was in preparing food for livestock. No analysis by power capacity was possible.

Finally we can note two negative pieces of evidence that suggest for earlier years a lack of use of gas engines, paradoxically in those areas that used most in 1925. For a number of years the RASE held an annual competition to assess the progressive nature of farms in different parts of the country. A number of farms in a region were visited on a number of occasions and their facilities and working assessed. A prize was awarded to the best and the judges provided a report for the Society's journal. That for the competition in Northumberland and Durham in 1887 notes that 'almost universally there was a fixed threshing machine driven either by steam or water power', with the engine also driving barn machinery.[42] A survey of typical farms in Cheshire and North Wales six years later indicates that farms that had invested in improvements had steam engines—others used horse or water power.[43]

AWARENESS

That farmers could be fully aware of the possibilities of newer forms of motive power is not in doubt. The primary means of disseminating information about agricultural innovations was via the various societies. These ranged from the national ones like the Royal Agricultural Society of England and the Highland and Agricultural Society of Scotland to the smaller local societies and clubs such as (taking the north-east at the end of the nineteenth century as an example) Bishop Auckland Agricultural Society, Malton Agricultural Society, Newcastle Farmers Club, Northumberland Agricultural Society, Ryedale and Pickering Agricultural Society and Yorkshire Agricultural Society.[44] It is true that these were run by people with influence in other areas of life—the RASE had on its committee MPs and two successive chairmen of the North Eastern Railway[45]— and the local societies tended to be clubs for the major landowners of the county; bankers, businessmen and gentry. Indeed, 'the more acres a man farmed the more likely he was to be a member of ... a society'.[46] Similarly the interests of some could be narrow; the Newcastle Farmers Club journal from 1879 to 1925 shows a complete lack of concern with implements and machines. On the other hand, the activities of the societies were available to all through the journals many produced and, most importantly, through their annual shows. Then, as now, agricultural shows were a mixture of competitions for livestock and produce and exhibitions of equipment. Moreover there were often equipment competitions especially by the larger societies where available models would be assessed, as in the trials of suction gas plant previously noted. So equipment could be evaluated as well as brought to the attention of potential users. Gas engines were exhibited for around 50 years at the RASE show (Table 4 gives a selection). With good weather, a three-day show would normally attract well over 100,000 visitors—there were 218,000 at the Royal Agricultural Society of England's show at Manchester in 1897[47]—so that potential customers had ample opportunity to see what was available. Clearly these shows were to a large extent intended for the general public and, one suspects, for non-agricultural equipment users as well. It seems unlikely, for example, that the National Gas Company exhibited a 85 bhp engine with suction plant at the RASE show of 1908 in Newcastle in the hope of selling many to farmers, whereas local heavy industry might be interested, landowners sometimes being the directors of engineering companies. We cannot, of course, make any assessment of the effect of shows on the purchase of equipment. The landowners might have the money to spend but no willingness, whereas their tenants might wish to mechanize but have little capital.

TABLE 4   Examples of gas engines exhibited at agricultural shows
(Royal Agricultural Society of England unless stated)

1877	Liverpool	Crossley-Otto
1878	Bristol	Simon's patent Eclipse
		Thomson Sterne
1879	Kilburn	Clerk's 2 stroke
1880	Islington	
	(Smithfield Club)	Robson's 2 stroke
1881	Islington	Crossley-Otto
	(Smithfield Club)	London Gas Engine Co.
1891	Doncaster	Tangye
		Fielding and Platt's first engine
1892	Warwick	Andrew 3 engines
		Crossley 6 engines from 3–88 hp
		Tangye 7 engines 2–16 hp
		Wells $\frac{1}{4}$ hp
1897	Manchester	Campbell 5 engines $2\frac{1}{4}$–35 bhp
		Crossley 6 engines 1–30 bhp, 2 with dynamos
		Hardy and Padmore 5 engines 1 manpower—
		$1\frac{1}{2}$ bhp, 1 with dynamo
		National range 1–14 hp
		Robey $13\frac{1}{2}$ bhp
		Tangye 6, 25 nhp
		Trusty $\frac{1}{2}$, 10 nhp
1908	Newcastle	Hornsby-Stockport with suction gas plant
		Crossley 4–74 hp
		Campbell 12, 25 bhp with magneto electric
		ignition
		Bates 20 bhp and suction gas plant
		National $14\frac{1}{2}$, 85 bhp
		Tangye 43 bhp and suction gas plant
1920	Darlington	Crossley 6–31 bhp
		National $4\frac{1}{2}$, 18 bhp
		Ruston & Hornsby 22 bhp with suction gas
		plant
1923	Newcastle	Crossley $6\frac{3}{4}$–34 bhp with suction gas plant—
		for lighting
		Fielding & Platt 35 bhp with suction gas plant
		Ruston & Hornsby 37 bhp with suction gas
		plant

(Data from Donkin, as note 10; *Journal, Royal Agricultural Society of England*,
1877–92; Royal Agricultural Society of England exhibition catalogues 1897–
1923.)

We should also note, if only in passing, that equipment manufacturers seemed to have a good network of agents to supply and maintain their products. The Crossley-Otto engine exhibited in Liverpool in 1877 was on the stand of their agent, so if Crossley recognized the need for an agent only twenty or so miles from their works it is reasonable to assume they had many more, especially as they had branch offices in London, Glasgow and Newcastle. Similarly, Priestman's petrol engine of 1888 was available that year from an agent in Newcastle[48] and later, in the 1920s, Blackstone petrol engines were sold by agricultural engineers in villages in Northumberland as small as Newton and Wall.[49]

APPLICATIONS

Gas engines were used in agriculture mainly to power what is conveniently called barn machinery, that is small static machines usually found in

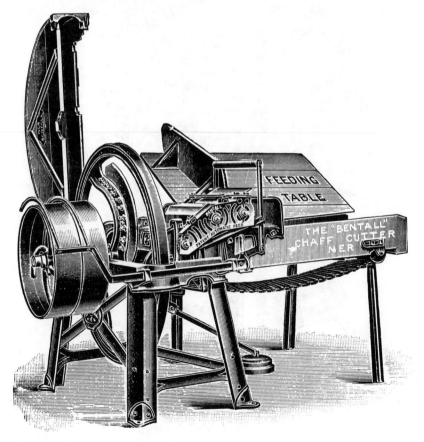

Figure 4. Chaff cutter (from D.N. McHardy, *Modern Farm Machinery*, Methuen, 1924).

barns or other farm buildings rather than used in the field. Such equipment was employed mostly for animal-food preparation, generally to grind and cut fodder into small portions to make it more digestible or to mix with other food. McConnell in his practical handbook of 1904[50] suggested that a mixed husbandry farm of over 100 acres would need a chaff cutter (to cut up straw), a turnip cutter, a winnowing machine (to separate grain from weeds and chaff), an oat bruiser and an oilcake breaker. Farms of over 500 acres would need an extra turnip cutter. As has already been noted, advertisements for engines tended to emphasize their use with chaff cutters but the Crossley list of uses of 1888[51] includes turnip cutting, wheat grinding, threshing, churning, straw cutting, milk separating as well as non-food applications like sawing, pumping and hoisting. Table 5 gives some typical features.

Figure 5. Oilcake breaker (from D.N. McHardy, *Modern Farm Machinery*, Methuen, 1924).

TABLE 5   Power requirements, output and costs of
barn machinery *c.* 1920

	bhp	*Output (cwt/hr)*	*Cost*
Thresher	12	18	£75–150
Grist-mill	4	4–5	£15–25
Oat crusher	3	6–12	£20–60
Chaff cutter	4	20	£15–35
Cake breaker	2	25–35	£10–15
Root cutter	1½		£25–35

(Data extracted from J.R. Bond, *Farm implements and machinery*, Benn Bros., 1923, ch. 19–21; Royal Agricultural Society of England exhibition catalogues 1920, 1923.)

The 1925 census gave a detailed breakdown of the uses of all the forms of motive power.[52] What is particularly clear is that with the exception of wind power, each engine was used for a number of purposes, though such diversity was less apparent for steam power than for others: 73 per cent of gas engines were used for 'chaff cutting and hay chaffing'; 35 per cent for 'turnip cutting, root pulping and cleaning', 47 per cent for 'milling and grinding' and 17 per cent for 'corn and cake crushing'. These proportions are broadly similar to those for oil, petrol and electric engines. Water power was used nearly as much for chaff cutting and milling but less for the other tasks and was also widely used for threshing. Over 60 per cent of steam engines were used for threshing, whereas relatively few electric or internal combustion engines were. That gas engines were usually stationary was limiting as threshing had become a field operation suited to portable steam engines and there was presumably no desire to return to older methods. This dominance of threshing work by steam engines appears to provide further evidence that gas and oil engines were introduced largely to complement the work of steam engines and not to replace them. Other uses for the gas engine listed in this census report are (in decreasing order of popularity) water pumping, sawing, lighting, elevating, kibbling (cracking maize for poultry food) and butter churning, these last three being represented by only a few engines.

COSTS

Looking at the purchase costs of gas engines can give some idea of the attraction they held, for engines could be very cheap. Gas engines were manufactured in a wide range of powers from 1 manpower to the massive twin-cylinder engines generating hundreds of horsepower, though those in use in agriculture probably never exceeded 20 bhp if, indeed, they were ever that powerful. In 1879, Crossleys could supply an Otto-type engine of ½ nhp for £60. Around this time a Bisschop engine cost £25 for 1 manpower and £50 for 4 manpower.[53] Crossleys' smallest was a ¾ ihp vertical engine costing £32 in 1888. A more typical size might be their 4 nhp (8 ihp) horizontal engine at £170. (A 5 hp Priestman oil engine at that time cost £235.)[54] Following the expiration of Otto's patent, the wide range of 4-stroke engines then available brought down prices and Crossleys were forced to reduce theirs immediately by a third. Thus in 1897 Crossleys' 4 nhp engine cost only £92 and Hardy and Padmore had a 1 manpower engine for only £7 17s 6d. This was at a time when the average wage in agriculture in England was 14s[55] though varying from 12s 7d in the eastern counties to as much as 18s in Northumberland and Derbyshire, so an engine could be bought for the equivalent of about three months' wages for a labourer. By 1902 a 12½ bhp Crossley engine with dynamo was £106 and a few years later Bates were offering a 20 bhp with suction gas plant for £200. Twenty years before, that price would have bought only 12 ihp (maybe 10 bhp) working from town gas. Advertisements in the 1870s and 1880s make reference to deferred payment schemes so clearly manufacturers (as now) were keen to encourage sales when available capital was not plentiful.

Reference has already been made to running costs in relation to the different forms of gas supply, but it is necessary to take some account of non-fuel costs and to compare costs of other forms of power.

The Otto–Langen engine was guaranteed to consume less than a penny-worth of gas/hp/hour and some costs of 1872 show it to be impressively cheaper than steam.[56] For a 1 hp engine in a printing works, the fuel, labour and maintenance costs for steam were £109 and for gas £36. It is not stated for which part of the country these figures apply but if it were one where coal was expensive then, given that £76 of the steam engine costs were for coal, the difference could be less marked, though still significant, in the cheaper areas. Consumption for the Otto–Langen engine was stated to be 26½ ft³/effective hp/hr.[57] Figures quoted a few years later comparing steam, gas (presumably Otto) and hot air engines of 12 hp were quite different: £155, £86 and £34 respectively.[58] These, however, were for fuel costs only and were based on use at full power for 10 hours/day. The intermittent use found in agriculture (where 12 hp would be too large an engine) would affect these estimates as, unlike the gas engine, the steam engine would still consume fuel even when not doing any work. Similarly, working at less than full power would reduce consumption. The hot air engine (in this case the Buckett type) seems especially cheap, but its disadvantages of bulkiness, low thermal efficiency and air leakage prevented it from being widely used. Other types of hot air engine could consume up to three times as much fuel.[59]

The popular Bisschop engine seems from advertisements[60] to have had costs similar to the Otto–Langen: the 1 manpower consumed 12 ft³/hr (for ½d) and the 4 manpower 28 ft³ at 1⅛d. (Incidentally, manufacturer claims for gas consumption are not always borne out by tests—one Bisschop engine consumed no less than 139 ft³/hp/hr though usually test figures were nearer to those quoted even if still greater.[61]) In their literature of the 1880s, Crossleys were claiming their smaller engines to be more economic than steam engines, unless work was intermittent, but noted the reduction in labour possible, as well as other advantages.

With gas plant, fuel costs were considerably lower, down to as little as 0·05d/bhp/hr for suction gas plant.[62] Oil engines at the same time were costing around 0·75d/actual hp/hr for a 12 bhp engine with higher costs for smaller ones.[63]

It is not easy to estimate total costs for the use of gas engines especially when use was intermittent. However, if we assume 5 hours' use per day of a 5 bhp engine and accept McConnell's figures of annual depreciation and wear and tear of 6 per cent[64] (he simply quotes for 'engines' without specifying type), then in the 1900s an engine costing £60 running on town gas at 1d/bhp/hr might cost something of the order of £39 per year plus labour costs. This cost is similar to the annual labourer's wage of the time, but in theory the engine had the capacity of twenty men.

Conclusions

Before drawing conclusions it is as well to remember the general condition of agriculture in Britain in the last quarter of the nineteenth century when the gas engine was an appropriate source of power. This was the time of the agricultural depressions when the lower prices of imported food had a disastrous effect in many parts of the country in a period when both floods and droughts occurred to make things even more difficult. There was a growth in intensified and specialist farming—such as market gardens and orchards—with different equipment needs. A decrease in England in the acreage of corn from 7·5 million in 1875 to 5·7 million in 1895 meant there was less corn to thresh or chaff to cut.[65] There was some increase in cattle from 4·2 million in 1875 to 4·7 million in 1885 but with a decrease to 4·5 million by 1895, hardly enough to make significant differences in the need for power for fodder preparation. With the number of sheep also falling, from 19·1 million in 1875 to 15·5 million in 1895, there was clearly less demand for the sort of work the gas engine could do with barn machinery. As in general there was less money available to spend on improvements, it is perhaps not particularly surprising that the gas engine was not so widely used as one might at first expect. Moreover, had the later more favourable economic conditions not coincided with the development of the more convenient oil engine, then use of the gas engine might have been greater.

Given the limitations of our sources, conclusions cannot be as definite as we would like, but the claims made in the introduction are borne out. Gas engines were widely available to farmers and offered a greater range of power outputs and prices than was possible with steam engines. Mechanical power of as little as $\frac{1}{4}$ hp was available at reasonable initial and running costs. That the gas engine's major disadvantage—its dependence, in its early years, on a fixed fuel supply—rendered it less competitive after the oil engine was introduced, clearly limited its widespread use in agriculture, though not in other industries. In this application, then, perhaps it is right to view the gas engine in the broader context of the adoption of internal combustion engines, for gas engines must have provided the initial stimulus to the wider use of mechanical power on farms even though this use was characterized much more by oil engines. That it was a wider use seems clear: farmers were not to any significant extent replacing steam engines by gas and oil but replacing horse and manpower by mechanical power. On the other hand, as we have seen, the gas engine did not disappear from use in the 1890s when the oil engine became popular. From the relatively static figures shown by the Agricultural Census, we might infer that the engines reported were merely ones that had been in use for a great many years, had performed satisfactorily and so had not been replaced by oil and that gas engines were archaic by then. However, when agricultural textbooks as late as the 1920s treated them still as a viable source of power and manufacturers exhibited at the Royal Agricultural Society of England's shows for half a century, then we might suspect that sales—if at a low level—continued long after the oil engine might have been expected to have become paramount.

The gas engine's role in agriculture might thus be characterized directly as having provided a small amount of the mechanical power used in the late nineteenth and early twentieth centuries and indirectly, and more importantly, as stimulating in a general manner the greater use of such power.

Acknowledgement

In its original form this paper was prepared as a dissertation for a CNAA Post Graduate Diploma in the Development of Science and Technology at Newcastle upon Tyne Polytechnic and I am grateful for the help of my supervisor, Joe Clarke. For less readily available source material I made use of the Hunday National Tractor and Farm Museum, Newton, Stocksfield, Northumberland for both artefacts and printed material.

Notes

Note: *JRASE* is *Journal of the Royal Agricultural Society of England*

1. K.A. Barlow, *A history of gas engines 1791–1900*, unpublished PhD thesis, University of Manchester, 1979, p. 318 and Fig. 1.1

2. Ministry of Agriculture and Fisheries, *The Agricultural Output of England and Wales 1925*, HMSO, 1927 (Cmd 2815), p. 125 (reproduced as Table 3).

3. J. Speir, 'Changes in farm implements since 1890' in *Transactions—Highland and Agricultural Society of Scotland*, Vol. 18, 1906, pp. 46–62.

4. N. Harvey, *The industrial archaeology of farming in England and Wales*, Batsford, 1980, p. 125.

5. H. Evershed, 'On the wear and tear of agricultural steam engines and threshing machines, whether fixed or portable' in *JRASE*, Vol. 23, 1862, pp. 323–38.

6. E.J.T. Collins, 'The age of machinery' in G.E. Mingay, *The Victorian countryside*, Routledge & Kegan Paul, 1981, Vol. 1, pp. 200–13

7. P. McConnell, *Notebook of agricultural facts and figures for farmers and farm students*, 7th edn, Crosby Lockwood, 1904.

8. Quoted in Collins, as note 6.

9. Board of Agriculture and Fisheries, *The agricultural output of Great Britain*, HMSO, 1912 (Cd 6277), p. 62 (reproduced as Table 2).

10. The outline is based mainly on Barlow, *A history of gas engines*; C.L. Cummins, *Internal fire*, Carnot Press, 1976; and B. Donkin, *A textbook on gas oil and air engines*, Griffin, 1894.

11. J. Robson, *A brief memoir of James Robson*, 1915.

12. J.R. Bond, *Farm implements and machinery*, Benn Bros., 1923, p. 207.

13. This section is based on Barlow, *A history of gas engines* and Donkin, *A textbook on gas oil and air engines*.

14. Crossley Bros., *The Otto gas engine*, 1888 (reproduced in Barlow, as note 1, Vol. 2, pp. 62-6).

15. Dowson Economic Gas & Power Company, *Dowson economic gas for driving gas engines and for heating purposes*, 1888 (reproduced in Barlow, as note 1, Vol. 2, pp. 166-71).

16. Crossley Bros., *Selected list of users of Crossley's 'Otto' gas engine driving appliances in connection with corn grinding, fodder chopping and similar work only*, c. 1888 (reproduced in Barlow, as note 1, Vol. 2, pp. 98-106).

17. W.J. Malden, 'Motive power of the farm' in *Transactions—Highland and Agricultural Society of Scotland*, Vol. 8, 1896, pp. 179-213. ·

18. J.E. Dowson and A.T. Larter, *Producer gas*, Longmans, Green, 1906; P.W. Robson, *Power gas producers*, Edward Arnold, 1908.

19. H.W. Buddicom, 'Miscellaneous implements exhibited at the Cambridge Show' in *JRASE*, Vol. 83, 1922, pp. 164-71.

20. H.W. Buddicom, 'Miscellaneous implements exhibited at Doncaster 1912' in *JRASE*, Vol. 73, 1912, pp. 195-205.

21. J.A. Voelcker, 'Annual report for 1919 of the consulting chemist' in *JRASE*, Vol. 80, 1919, pp. 396-406.

22. J. Middleton and R. Stanfield, 'Trials of suction gas-producer plants' in *Transactions—Highland and Agricultural Society of Scotland*, Vol. 18, 1906, pp. 208-24.

23. H.R. Sankey, 'The trials of suction gas plant at Derby 1906' in *JRASE*, Vol. 67, 1906, pp. 154-84.

24. Royal Agricultural Society exhibition catalogue.

25. Speir, *THASS*, 1906.

26. Bond, as note 12, p. 207.

27. Calculated from the figures in Table 2.

28. D. Pidgeon, 'Report on miscellaneous implements at Newcastle' in *JRASE*, Vol. 24, 1888, pp. 195-216.

29. D. Pidgeon, 'The evolution of agricultural implements' in *JRASE*, Vol. 3, 1892, pp. 49-70, pp. 238-58.

30. Crossley Bros., as note 14.

31. Donkin, as note 10, p. 2.

32. Malden, as note 17.

33. Bond, as note 12, p. 207.

34. Leaflet in Northumberland County Record Office (NRO 309/H1/2).

35. Donkin, as note 10, p. 297.

36. Barlow, as note 1, p. 348.

37. Reproduced in Barlow, as note 1, Fig. 5.21.

38. Reproduced in Barlow, as note 1, Fig. 5.28.

39. Reproduced in Barlow, as note 1, Fig. 7.12.

40. Crossley Bros., as note 16.

41. D.K. Britton and I.F. Keith, 'A note on the statistics of farm power supplies in Great Britain' in *Farm economist*, Vol. 6, 1950, pp. 163–70.

42. W.C. Little, 'Report on the farm prize competition in Northumberland and Durham in 1887' in *JRASE*, Vol. 23, 1887, pp. 582–660.

43. J. Bowen-Jones, 'Typical farms in Cheshire and North Wales' in *JRASE*, Vol. 4, 1893, pp. 571–620.

44. K. Hudson, *Patriotism with profit: British agricultural societies in the eighteenth and nineteenth centuries*, Hugh Evelyn, 1972, p. 134.

45. J.A.S. Watson, *History of the Royal Agricultural Society of England, 1839–1939*, c. 1940.

46. Hudson, as note 44, p. 100.

47. N. Goddard, 'Agricultural societies' in G.E. Mingay, *The Victorian countryside*, Vol. 1, pp. 245–59.

48. Leaflet in Northumberland County Record Office (NRO 309/H1/2).

49. Leaflet in Northumberland County Record Office (NRO 322.2 Box 8).

50. McConnell, as note 7, p. 60.

51. Crossley Bros, as note16.

52. Ministry of Agriculture and Fisheries, *Agricultural output*, p. 127.

53. Prices in this section extracted from lists reproduced in Barlow, as note 1, and from Royal Agricultural Society exhibition catalogues.

54. W.C. Unwin, 'Trials of light portable motors at Plymouth' in *JRASE*, Vol. 25, 1890, pp. 580–603.

55. McConnell, as note 7, p. 440.

56. Barlow, as note 1, p. 164.

57. Advertisement by Sleicher, Schumm & Co., Philadelphia (reproduced in Barlow, as note 1, Fig. 5.21).

58. J. Scott, *The complete textbook of farm engineering*, Crosby Lockwood, 1885, Part 5, Barn implements and machines, p. 49.

59. Donkin, as note 10, Part 3.

60. Reproduced in Barlow, as note 1, Fig. 5.28.

61. Donkin, as note 10, Appendix, Section E.

62. Middleton and Stanfield, as note 22.

63. McConnell, as note 7, p. 44.

64. McConnell, as note 7, p. 58.

65. Figures extracted from C.S. Orwin and E.H. Whetham, *History of British agriculture 1846–1914*, rev. edn, David & Charles, 1971, Chapters 9 and 10.

The Contributors

P.S. BARDELL is Senior Lecturer in Mechanical Engineering at the Colchester Institute of Higher Education, Essex. He is currently engaged on doctoral research into the history of railway locomotive balancing.

MARJORIE N. BOYER is Professor Emerita of History at York College of the City University of New York. She is the author of *Medieval French Bridges*.

MICHAEL DUFFY is Senior Lecturer in Mechanical Engineering at Sunderland Polytechnic; his particular research interest is the history of railway engineering.

JOSÉ A. GARCÍA-DIEGO has a professional engineering practice in Madrid and has recently (1983) edited and published the *Los Veintiun Libros de los Ingenios y de las Maquinas* which traditionally has been attributed to Juanelo Turriano.

DONALD R. HILL is an authority on Islamic technology who has translated and annotated the works of al-Jazarī and the Banū Mūsà.

ROBERT J. SPAIN is a professional engineer with the Greater London Council and is currently carrying out doctoral research into the history of Roman water-mills.

IAN R. WINSHIP is a science and technology librarian at Newcastle-upon Tyne Polytechnic and is currently preparing an MA thesis on some mechanical aspects of railway safety in nineteenth-century Britain.

Contents of Former Volumes

Third Annual Volume, 1978

Fourth Annual Volume, 1979

Fifth Annual Volume, 1980

D.G. TUCKER, Emile Lamm's Self-Propelled Tramcars 1870–72 and the Evolution of the Fireless Locomotive.

S.R. BROADBRIDGE, British Industry in 1767: Extracts from a Travel Journal of Joseph Banks.

RICHARD L. HILLS, Water, Stampers and Paper in the Auvergne: A Medieval Tradition.

Sixth Annual Volume, 1981

MARJORIE NICE BOYER, Moving Ahead with the Fifteenth Century: New Ideas in Bridge Construction at Orleans.

ANDRÉ WEGENER SLEESWYK, Hand-Cranking in Egyptian Antiquity.

CHARLES SÜSSKIND, The Invention of Computed Tomography.

RICHARD L. HILLS, Early Locomotive Building near Manchester.

L.L. COATSWORTH, B.I. KRONBERG and M.C. USSELMAN, The Artefact as Historical Document. Part 1: The Fine Platinum Wires of W.H. Wollaston.

A. RUPERT HALL and N.C. RUSSELL, What about the Fulling-Mill?

MICHAEL FORES, *Technik*: Or Mumford Reconsidered.

Seventh Annual Volume, 1982

MARJORIE NICE BOYER, Water Mills: a Problem for the Bridges and Boats of Medieval France.

WM. DAVID COMPTON, Internal-combustion Engines and their Fuel: a Preliminary Exploration of Technological Interplay.

F.T. EVANS, Wood Since the Industrial Revolution: a Strategic Retreat?

MICHAEL FORES, Francis Bacon and the Myth of Industrial Science.

D.G. TUCKER, The Purpose and Principles of Research in an Electrical Manufacturing Business of Moderate Size, as Stated by J.A. Crabtree in 1930.

ROMAN MALINOWSKI, Ancient Mortars and Concretes: Aspects of their Durability.